Contemporary Islam

Merryl Wyn Davies series editor

Science and Muslim Societies

Science and Muslim Societies

Nasim Butt

Grey Seal London

First published 1991 by *Grey Seal Books*
28 Burgoyne Road, London N4 1AD, England

Copyright 1991 Nasim Butt

All rights reserved. No part of this publication may be reproduced or transmitted in any form or by any means, electronic or mechanical, including photocopy, recording or any information storage or retrieval system, without permission in writing from the publishers or their appointed agents.

British Library Cataloguing in Publication
  Butt, Nasim
    Science and Muslim societies.- (Contemporary Islam)
    I. Title II. Series
    297.1975
    ISBN 1-85640-023-9

The author Nasim Butt is Head of Science at King Fahad Academy, London, and Educational Consultant for Science to the Iqra Trust. After earning the BSc in food science and technology, Butt moved to the history and philosophy of science for his MSc at University College, London. He then obtained the PGCE so that he could teach in British schools, before beginning a part-time PhD course at the University of London's Institute of Education. He has combined the teaching of chemistry and general science to secondary school students through A-level with religious studies (Islam), and during the past few years has concentrated in his research and writing on the impact of the British science national curriculum for Muslim students and teachers.

Contents

Preface	vii
Introduction	1
1 The Roots and Nature of Western Science	10
2 Islamic Science: The New Paradigm	37
3 Islamic Science in History	65
4 Islam and Science Education	89
Bibliography	108
Appendix 1 Science Attainment Targets	115
Appendix 2 Muslim Scientists	119
Index	130

Preface

The central aim of this book is to popularize the concept of Islamic science and impart a general awareness of its ethical parameters and conceptual matrix. Along the way, some of the glorious aspects of the history of Islamic science are highlighted, and science education is discussed in the context of the British National Curriculum.

Much of the ideas developed in the last chapter on science education will also be applicable to the colleges and universities of the Muslim world. The educational system there has bifurcated into the religious and the secular, a division encouraged by the government of those countries. Secular education, of course, has the greater influence on students because it is practised most of the time. In essence, Islam has been marginalized from science education, rather than being the central plank on which science rests and from which it draws its continuous inspiration.

A final word of thanks to Ziauddin Sardar who encouraged me, and to my wife and daughter who allowed me to write this book. Any errors which remain, of course, factual or otherwise, are solely mine. May Allah grant me the strength and perseverance to continue to produce work for His sake only.

Dedicated with love
to my mother, Munawar Sultana,
and my late father, Mohammed Sadiq Butt

Introduction

Whether beneficial or otherwise, the fruits of modern science--from the steam engines and machinery of the industrial revolution, the chemical weapons of World War I and the atomic bombs of World War II to today's microchip technology and ethically questionable techniques of genetic engineering--are arriving, often unheralded, in a society that may be unprepared and has to discover how to respond.

Some of the impacts of the scientific enterprise are more conceptual than practical and technological. Examples are the displacement of the sun-centred picture of the solar system by the geocentric theory, and the Darwinian evolutionary theory, both of which had a profound influence on Western man's conception of his place in the universe. There are also the 'accidental' impacts of modern science, such as the explosion at the Bhopal chemical factory in India in December 1984, or that at the Chernobyl nuclear plant in the Soviet Union in April 1986. Damage caused to wildlife by pesticides, as first highlighted by Rachel Carson in her eloquently-written *Silent Spring* (1962), and the excessive use of animals for vivisection and product research are also constant sources of consternation.

The industrially underdeveloped world is not insulated from these effects either. Peasant communities were promised freedom from famine and social disruption by the 'green revolution'. Satellite dishes have reached remote villages for purveying educational enlightenment or subversive propaganda. Western technology, under the guise of modernization and development has fostered an illusion of achieving inudstrial competence, measuring the worth of a nation through the Gross Domestic Product (GDP) and other economic indicators.

It appears, therefore, that scientific research is like a Pandora's box, spawning products which society may or may not welcome. Individual members of society, however, have little influence over the genesis and deployment of these products. The real power lies with the institutions and their paymasters who decide on the aims and objectives of research and research priorities, thereby ultimately dictating the direction in which society goes.

Public opinion is powerful. Common thoughts and emotions are rapidly disseminated through society, which has been the cause of misconceptions about the status of science. Many people think that science deals only with things which can be 'proved scientifically', a phrase often used as though there were no need for subsequent argument. One of the main features of modern science since the 1930s is the realization that there are no absolute certainties in science. Scientists have to start with some kind of hypothesis or framework which predicts results that subsequently either support it or show it to be deficient in some respects. If the latter is the case, then a new improved theory is advanced, which in its turn predicts further results from experiments. Admittedly, scientists can be dogmatic, tenaciously holding on to some brainchild theory unless and until empirical evidence is so decisively against it that if they did not succumb they would risk ostracism from the scientific community. What essentially needs to be stressed is that at no stage can a scientist claim a theory to be absolutely true or absolutely proved.

Another fundamental feature of science is that regardless of the number of experiments found to be successful, if one

Introduction

correctly-performed test shows the theory predicting an incorrect result, the whole theoretical edifice falls. One new fact repeatedly shown to be correct can invalidate a centuries-old theory. No theory, therefore, can guarantee that such a decisive test will not be devised to invalidate it. Such uncertainty reinforces the provisional nature of scientific theories. The task of science is now to devise tests designed to determine the limits of applicability of theories rather than their absolute verification.

The distorted picture of the past four centuries of science as a search for absolute truth, has now been drastically revised. Unfortunately, however, many scientists still give the impression that science intrinsically and by itself, has the ability to solve most global problems. Yet when it comes to the question of the destructive impact of science on society, they claim they are not culpable and free themselves of moral responsibility. It is sanctimoniously claimed that any metaphysical or theological parameters are outside the realms of science and therefore have no part to play in scientific research.

One example shows the naivete (if naivete it is) of such claims. Scientists operate on the principle of simplicity when it comes to choosing between two theories that adequately explain the same phenomena equally. The principle, known as 'Ockham's razor' after William of Ockham (1280-1349) who first propounded it, states that 'multiplicity must not be postulated without necessity'. This basis for assuming natural laws are simple means that of two explanations the simpler is preferable regardless of empirical criteria as to why this should be so.

In the total eclipse which took place on 25 May 1919, Eddington measured the shift of light from a star as it came past the darkened sun. There were thought to be two ways of interpreting the results. Either they

Could be accounted for by a slight modification of Newton's

> law of gravitation in which the index in the law was made slightly different from z. A law with an index slightly different from an integer is in reality an extraordinarily complicated one, and has therefore so small a probability that it does not merit serious consideration.
>
> Jeffreys

Einstein's theory of relativity provided an alternative explanation. Although intellectually daunting, it was found to be simple in statement, a feature which Jeffreys called the 'simplicity postulate'. The point to note is that there is no empirical proof that the postulate is correct or true, but because it has a powerful aesthetic appeal it 'feels right' and remains unquestioned.

The belief that there are no assumptions in science and that, therefore, everything can be 'proved scientifically' is one of the more persuasive errors about science. In fact, all science rests not on self-evident but on unprovable axioms. To understand what this means one has only to ask any nuclear physicist what is an atom, an electron or a quark. He or she will be unable to explain clearly and unambiguously the nature of an elementary particle or entity because, in reality, it is just not known. Recently I attended a conference on particle physics for sixth formers at the Royal Institution, London. During questiontime a professor of physics from an eminent university, who is due to have a newly discovered particle named after him, was asked a deceptively straightforward question by one of the students. The questioner wanted to know what exactly we mean by the descriptive entity called 'charge' and why it is of an opposite value for elementary particles and their anti-particle counterparts. After deliberating for thirty seconds or so, the professor said that the question was difficult to answer beyond saying that charge is one important property which characterizes an elementary particle.

The scientist therefore uses what knowledge is already available to get on with the business of living and

Introduction

constructing things with partial knowledge. We see, for example, that electronic equipment is everywhere. Television tubes use electrons, and microchip calculators rely on advanced circuitry. Enough is known of the properties of electrons to enable equipment to be designed and made to work, yet no one has seen an electron. Everything we know about electrons is from the traces they leave in other media. The charge and mass of the electron have been derived as a result of the behaviour of the charge in a magnetic and electric field. Because an electron beam produces a diffraction pattern on a fluorescent screen, the electron can also be described as having a wavelength. The point is that different models are used for different applications. It is difficult to comprehend that an object can have mass, charge and wavelength all simultaneously. In the design of an electron microscope, the electron beam is considered as having a wavelength. For the functioning of a television tube the particles have charge and mass. Attention is usually not drawn to the scientist's inconsistency in describing a given object in ways which are intellectually incompatible. Such is the blind faith people have in the scientific enterprise.

It can safely be said that much influential, but unfounded, ideology of our time involves an extension of science well beyond its legitimate limits, so that all problems--social, political, economic, environmental--are construed as scientific ones. Religion is deliberately marginalized as an intellectual irrelevancy and solutions to societal problems offered in a way that obscures the social and political issues at stake. The pseudo-sciences of sociobiology and social Darwinism, for example, pose as explanations of social phenomena, thereby disguising the political realities and serving to justify such various kinds of oppression as that of the poor, women or racial minorities.

The gulf between rich and poor and between developed and underdeveloped countries widens consistently. The environment is being destroyed and the threat of annihilation

Science and Muslim Societies

looms. The social and political problems facing us, as reflected in our society and educational institutions, are urgent and vital. This is not in question. What needs to be assessed is whether the Islamic ethical system, based on a pristine monotheism, provides a comprehensive and effective antidote to the ideologization of science with its destructive impact on society. That such a system was successful in the past, and provides an ethical blueprint for the development of contemporary science and science education as well as an holistic alternative most in tune with human nature, is a line of thought assiduously developed throughout this book.

The idea of an Islamic alternative to Western science was proposed by Seyyed Hossein Nasr in *The Encounter of Man and Nature*. He argued that because the content and applications of Western science have become divorced from revelatory knowledge as a result of secularization, the whole enterprise is therefore illegitimate and highly dangerous. The basis of religious morality has been overturned so that man-made values are now used as the ultimate arbiters between right and wrong. In a most revealing passage, Nasr declares:

> The independent critical function which reason should exercise vis-a-vis science, which is its own creation, has disappeared so that this child of the human mind has itself become the judge of human values and the criterion of truth. In this process of reduction, in which the independent and critical role of philosophy has itself been surrendered to the edicts of modern science, it is often forgotten that the scientific revolution of the seventeenth century is itself based upon a particular philosophical position. It is not *the* science but *a* science making certain assumptions as to the nature of reality, time, space, matter, etc. But once these assumptions were made and a science came into being based upon them, they have been comfortably forgotten and the result of this science made to be the determining factor as to the true nature of reality.
>
> Nasr, *Encounter*

Introduction

This is the very point. The nature of reality can also be described from a non-Western intellectual perspective. In fact, some Muslim writers have attempted to construct a model of science based on Islamic thought and quranic conceptual analysis, which shows, sig-nificantly, that scientific activity can flourish in an environment where the views regarding nature, knowledge and methodology are different from those of the dominant views in Western culture. Islamic science has its characteristic conception of values, know-ledge and methodology which can provide a better social structure and ethical framework for the growth of science.

The holistic perspective which Islamic science provides poses a fundamental challenge to the reductionist methodology of Western science. Here issues of morality and values are marginalized, and only those aspects of a phenomenon which are amenable to pure reason are considered worthy of theoretical and practical investi-gation. Macroscopic phenomena are reduced to and explained in terms of microscopic processes: mental events in terms of the electro-physiological processes in the brain, mathematics in terms of logic, even social structures in terms of the relationships between the actions of individuals. This is not to say that a reduc-tive explanation in itself is a bad thing, but only when it is the only explanation. The multi-dimensional global ecological crisis is an eloquent testimony to the harm that methodological exclusivity (for which read reductionism) can do to the relationship between man and nature.

For these reasons, among others, there is a widespread interest in the return of Islamic science. Muslim societies are becoming more conscious of their traditional heritage and distinct cultural identities. The outcome of the present crisis in science and the intellectual awakening of the Muslim people have together resulted in calls for the contemporization of Islamic science. Science is obviously perceived to be the enterprise that, more than any other, can be used to ameliorate human conditions. There is great disillusionment, however, with Western science which is blamed for its assaults on humanity and the natural environment, thereby dehumanizing and alienating people. This dislocation can be effectively countered by the ethical, moral, and generally holistic, perspective of Islamic science.

The conceptual values on which Islamic science is built must therefore be brought into the classroom. They must be discussed openly and explicitly to strengthen the foundations of science edu-cation. In the British National Curriculum some attempt has been made to integrate science with religion. For a society based on the philosophy of secular humanism this is a commendable effort, al-though of course it still leaves much to be desired from the Islamic point of view. Notwithstanding the deficiencies, I believe that the National Curriculum is open to the introduction of the values of Islamic science into the classroom, and is consequently far better than the present educational system that has its roots in a colonial past and within the framework of which science is taught.

As far as Muslim societies are concerned, Islamic science needs to be integrated with and become part of our education system. From a civilizational perspective, science and technology are fundamental components of the Islamic worldview, which has the sharia and epistemology at its core from which flow fundamental conceptual and institutional structures: political and social organizations, economic enterprise, environment and, of course, science and technology. We can see, therefore, the fundamental relationship between science education and the central core, the ethics and methodology within which it operates and from where it draws its continuous inspiration. Science teachers and students, consequently, have a significant role to play in an Islamic society. If they are to fulfil their roles at all adequately, they must have a profound awareness of the

Introduction

ethical dimensions of knowledge and be able to look at the ethics of science critically. This point is further developed in my final chapter, firmly integrating science education with Islamic science and into Muslim societies. To effect such a synthesis is the central aim of this book. As I believe that present curricula in most Muslim countries are incredibly colonial in nature, the analysis presented in the final chapter is of relevance to Muslims in all parts of the world.

1

The Roots and Nature of Western Science

> If what we are discussing were a point of law or of the humanities, in which neither true nor false exists, one might trust in subtlety of mind and readiness of tongue and in the greater experience of the writers, and expect him who excelled in those things to make his reasoning more plausible, and one might judge it to be the best. But in the natural science, whose conclusions are true and necessary and have nothing to do with human will, one must take care not to place oneself in the defense of error; for here a thousand Demostheneses and a thousand Aristotles would be left in the lurch by every mediocre wit who happened to hit upon the truth for himself.
>
> Galileo Galilei

Such an absolutist view of science has been the hallmark of Western civilization for the last four centuries, ever since Galileo and his contemporaries propounded it. It is a worldview elaborated by Francis Bacon in the early seventeenth century. The essence of this so-called Baconian picture of the scienfific method is that scientific investigations start with the accumulation of open-minded observational data, followed by the development of a hypothesis aimed at explaining the data, and then the testing of the hypothesis by key experiments. Empirical verification of the hypothesis endows it with the status of a scientific law, becoming a permanent addition to the body of scientific (or certain) knowledge.

To give an example, suppose that a scientist wished to investigate the relationship between the pressure and volume of fixed masses of gases at constant temperatures (now known as Boyle's law). To begin with, he might carry out an extended series of experiments with gases subjected to different pressures, noting the changes in volume produced as a result. After the accumulation of what he feels to be sufficient data, the scientist would then look for some common feature of the various pressure-volume situations studied and attempt to formulate a general principle capable of comprehensively describing his observations. By this means he would (according to the Baconian view of science) discover the pressure-volume relationship for fixed masses of gases at a constant temperature. The formulated principle would then be used to make specific predictions about what would happen in pressure-volume situations not yet investigated, and the predictions tested by means of further experiments. The principle would be verified if the predictions proved to be correct, raising it to the esteemed level of a scientific law--a permanent addition to the body of scientific knowledge.

In the above process, individual observations leading to specific statements about the natural world are used to develop a general statement (the hypothesis). The process is known as 'induction' and is central to the Baconian picture of the scientific method. For four centuries it has been used as the criterion to distinguish science from non-science, having had a powerful hold on the minds of many eminent scientists and academics. This carefully articulated worldview laid down the following conditions that must be satisfied for generalizations such as 'metals expand when heated' to be considered scientifically legitimate:

1. The number of observation statements forming the basis of generalization must be large.
2. The observations must be repeated under a variety of

conditions.

3. No accepted observation statement should conflict with the derived universal law.

Clearly, it is not legitimate to conclude that all metals expand when heated on the basis of a single observation of a metal bar's expansion. Moreover, further to enhance scientific legitimacy, bars of different types of metal of different shapes and sizes would need to be heated over a range of temperatures and pressures. Only then would the stated conditions be adequately met.

Inductivism (or the Baconian picture of the scientific method) has certainly not been an ephemeral worldview. Doubts about its validity were raised by the eighteenth-century Scottish philosopher, David Hume, who showed that induction is logically inadmissable, since no general statement can ever be derived from a finite number of individual observations. One common example quoted to illustrate this point is the impossibility of proving the truth of a statement such as 'all swans are white'. No matter how many white swans are observed in the world, the possibility still exists that a black (or non-white) swan may be observed one day. Induction cannot be justified purely on logical grounds.

Bertrand Russell's amusing anecdote of the inductivist turkey graphically makes the point. On its first morning at the turkey farm, this turkey perceptively noted that it was fed at 9:00 A.M. Rather than jumping to unjustified conclusions, the turkey accumulated a large number of observations to the effect that it was fed at 9:00 A.M. every morning. Other variables were also assessed--different days of the week and different weather conditions--but no change in feeding time was noted. Being a good inductivist, the turkey thus concluded that it was always fed at 9:00 A.M. This inductive inference, however, was gruesomely shown to be false when on Christmas Eve, instead of being fed, the turkey had its throat slit. The inductivist appeal to logic had misfired.

The principle of induction has created great turmoil for those attempting to explain how science works. By casting doubt on induction, Hume cast doubt on the status of scientific knowledge as certain truth, since all scientific law was thought to be of the same form as the statement 'all swans are white' and was therefore incapable of being verified.

What lay behind Baconian method?

Francis Bacon had a worldview, a vision of society that would materialize through scientists adopting the intellectual method he advocated. In his writings he stressed that the acquisition of genuine knowledge of Nature is tantamount to the acquisition of power. As a result, our power to act, to do good, to transform the human condition immeasurably for the better would be enhanced. Progress in science was essential in order to achieve such radical human, social progress. This was to be achieved by means of organized inquiry which based its results firmly on an empirical base (observational experiment), and equally by firmly ignoring the speculative, prejudicial and mythical world of philosophers and ordinary morals.

Such ideas, so articulately expressed by Bacon, came to exert a powerful influence over the rise and subsequent development of modern science. The idea that organized inquiry is needed in order that knowledge may be progressively acquired was a source of inspiration for both individual scientists and scientific institutions. Such thinking inspired the founding of the Royal Society and the work of Charles Darwin who proudly claimed in his most illustrious book, *The Origin of the Species*, to have based his scientific investigations on Baconian principles. Central to these principles was the idea that knowledge is to be acquired by ignoring the speculations of philosophers and, instead, arriving at results based on observation and experiment alone. Such a view has dominated all subsequent science.

Rene Descartes's independently developed but astonishingly influential dualistic theory of mind and matter accommodated itself neatly into the Baconian paradigm. Cartesian dualism, as it is called, divides up reality into two sharply distinct worlds: on the one hand, the objective world of fact, matter, physical reality; on the other, the subjective world of mind, consciousness, personal experience, value. Once this view is accepted (as it was at one time by most Western scientists in one form or another) the belief that the practice of science necessitates a sharp split between fact and value, objective reality and subjective feelings and desires, is a natural corollary. Although Descartes (admittedly) held that reason as well as experience is a source of incontrovertible knowledge, it is an historical fact that his aprioristic methodology was influential for a time, on the Continent at least if not in England.

These were the ideas prevalent in the seventeenth century which saw the completion of the scientific revolution in Britain. God was gradually detached from the affairs of society, for with the scientific revolution came the realization that humankind could make progress, that the good life did not have to be of necessity a life after death but could be achieved on this earth. Utopian works abounded. The writers were convinced that wise use of the new knowledge would lead to societies free from disease, violence and brute labour, where human brotherhood and peace would reign supreme. Science, by providing the world with an understanding of nature, would be the intellectual vehicle for realizing the good life on earth.

To gain power over nature was to be one of the central aims of the new science. In the words of Bacon himself, to ensure 'the enlarging of the bounds of Human Empire', and to achieve 'the effecting of all things possible'. As such, knowledge of causes of phenomena was what was required, not knowledge of purposes and goals in Nature. Aristotle's belief in purposive behaviour in Nature was ridiculed by Bacon since it did not conform with man being a master and possessor of Nature. Descartes expressed Baconian sentiments in the following way:

> It is possible to obtain knowledge highly useful in life, and that in place of the speculative philosophy taught in the schools, we can have a practical philosophy by means of which, knowing the force and actions of fire, water, air, the stars, the heavens and all the other bodies that surround us as distantly as we know the various crafts of the artisans--we may in the same fashion employ them in all the uses for which they are suited, thus rendering ourselves masters and possessors of nature.
>
> Descartes, in Rossi, p. 104

In one important sense, Bacon's vision has been confirmed in the Western world during the nineteenth and twentieth centuries. The great industrial, technological and medical progress achieved in the West is intimately associated with scientific progress. Progress in knowledge has eclipsed factors such as faith, morality, tradition and justice. The view that real human progress could be achieved through progress in science and technology has been held tena-ciously. The core of this belief has been that academic inquiry should be devoted to the achievement of knowledge and that the scientific enterprise is the paradigm of rationality. The majority of twentieth-century philosophers of science, although highly critical of Baconian method, have hardly made an intellectual dent in this basic tenet.

Karl Popper and Scientific Method

It was Karl Popper who seriously tackled and attempted to come to terms with one of the central problems of philosophy for over two hundred years, the 'problem of induction' (as it came to be called). Its solution defied the analysis of some of the great Western philosophers of the eighteenth and nineteenth centuries, such as Immanuel Kant and Bertrand Russell. If the principle of induction were true, then scientific knowledge risked being fundamentally flawed. Popper proposed an alternative picture of science in which the problem of induction disappeared.

Popper overturned the commonly held notion of empirical verification of scientific statements in favour of their empirical refutation (falsification). In the language of the layman, although a statement such as 'all swans are white' can never be proved, it can be disproved by the observation of a single non-white swan. Thus, although it is logically impossible to demonstrate that a scientific law is universally true without carrying out an infinite number of observations, such a law can, in principle, be refuted by a single properly authenticated observation that does not fit in with its predictions. In other words, scientific laws should be tested not by attempting to prove them right, but by attempting (in the most rigorous manner possible) to prove them wrong. This principle is central to the Popperian view of science and is epitomized by Popper's '*principle of empiricism* which asserts that in science, only observation and experiment may decide upon the *acceptance or rejection* of scientific statements including laws and theories' (*Conjectures*, p. 54).

An immediate consequence of Popper's reversal of the traditional approach to scientific method was that scientific laws fell from the pedestal of certain knowledge to the ground of working hypotheses. Scepticism, doubt and criticism are key words in Popper's philosophical vocabulary. If a scientific law persistently fails to disagree with our observations, it should be replaced by a new law that fits all the facts currently available. For example, Newton's law of gravitation (accepted as 'certain knowledge' for over 200 years) was eventually replaced by Einstein's general theory of relativity. Scientific laws are temporary phenomena, and scientific knowledge is a continuously evolving activity rather than an ever-growing pyramid of certain truth.

According to Popper, the first stage of scientific development is recognition of the problem. A theory is then formulated to explain the problem, which is tested rigorously by empirical setups designed to falsify it. This is done by using the theory to make definite predictions and then checking these predictions by means of experiments. If the results prove to be in total agreement with the predictions of the theory, the latter can be adopted as a working hypothesis until such time as its predictions fail to be corroborated by future experiments. If the results appear to be in broad agreement with the predictions of the theory but differ in certain important respects such as exact numerical values of certain quantities, the theory can be modified slightly by changing its detailed structure or by extending its scope. In the most radical outcome, of course, the experimental results prove to be completely incompatible with the predictions of the theory, indicating that the theory is fundamentally wrong and should be abandoned. As an illustration of the Popperian method in action let us examine the work that was carried out in the field of particle physics between the early 1960s and mid-1970s.

Ever since the construction of increasingly powerful particle accelerators from the late 1940s onwards, the relatively simple picture of the fundamental constituents of the universe has become increasingly complicated. Fundamental particles, namely the proton, neutron and electron can no longer be called 'fundamental'. In the early 1960s, the American physicist Gell-Mann put forward the 'three-quark theory' to explain the exis-tence of the new particles discovered since the late 1940s. He said that the fundamental particles were actually composite particles made up of three basic particles known as quarks. The theory went further and made a number of falsifiable predictions, including that of the existence of an as yet unknown particle called the 'omega minus'. This particle was subsequently discovered by physicists working at the Brookhaven Laboratory in 1964. Gell-Mann's theory was triumphantly vindicated and adopted as the basic model underlying particle physics.

Among the important predictions rigorously tested and corroborated during the next decade was the heterogeneity of the proton. The theory indicated that the proton has an internal structure consisting of three quarks held together by a new force more powerful than that holding together the atomic nucleus. By the mid-1970s, experimental testing had confirmed Gell-Mann's prediction.

Where experiments produced results which did not fit in with the predictions of the three-quark theory, physicists extended the scope of the theory by postulating the existence of a fourth type of quark, the so-called 'charmed' quark. This enabled discrepancies between theory and experiment to be explained and a new set of falsifiable predictions to be made. Whether or not the current theory proves durable or is replaced by a better and more powerful theory remains to be seen.

Popper tried to formulate a universal, ahistorical account of scientific methodology. Falsification was his criterion of demarcation between science and non-science. If, however, we try to extract from Popper's writings falsificationist criteria, either for the acceptance or rejection of theories within a science or for designating whole areas as scientific or non-scientific, we run into serious problems. That is, if we make our falsificationist criteria too strong then many well-advanced theories would fail to qualify as good science while if we make them too weak few areas would fail to qualify.

If falsified theories are to be rejected, then Newton's astronomy for one would fail to meet the demand. Observations ranging from the moon's orbit to those of the orbit of the planet Mercury were incompatible with Newtonian astronomy. Matching Newtonian theory with the moon's orbit involved assumptions about the shape of the Moon and its internal motions, as well as those of the Earth, corrections of telescopic readings to allow for refraction in the Earth's atmosphere and so on. Newton's theory was eventually saved by locating the cause of apparent falsifications elsewhere. The orbit of Mercury, however, presented a formidable problem. Had nineteenth-century physicists been Popperian falsificationists, the Newtonian theory would have been refuted. As it happened, they proceeded to develop the Newtonian theory in spite of the unsolved problem of Mercury.

Despite these problems, the ideas of Karl Popper have influenced a whole generation of scientists, affecting not only the way in which they carry out their work, but also their entire attitude towards science. Before Popper, incompatibility of predictions with experimental results was a mark of scientific failure, and reputations could seriously suffer as a result. But Popper put forward the thesis that only when ideas are subjected to stringent testing and possible falsification can science advance at all. Despite its shortcomings, many Western scientists have accepted this advice.

Kuhn and the Structure of Scientific Revolutions

Thomas Kuhn, the American philosopher of science, has probably made the most significant contribution, after Popper, to contemporary Western philosophy of science. His most illustrious book, *The Structure of Scientific Revolutions*, had an immediate and profound impact on contemporary thought, causing an entire generation of Western scientists to look anew at their discipline.

One of the most important ideas developed by Kuhn is the concept of a paradigm which, as Kuhn himself points out, defies precise definition. It can, however, be taken to be the basic, generally-accepted theoretical model that underlies a particular branch or sub-branch of science at any given time. In physics, for example, the current paradigm is that the atomic nucleus consists of an aggregate of positively-charged protons and uncharged neutrons bound together by a powerful nuclear force. All physicists accept this paradigm without question and use it as the basis of all their work. Indeed, it is the fact that their work is effectively based on the same paradigm and, as a result, is governed by the same basic procedures, rules and standards that identifies them as 'physicists'.

According to Kuhn, all branches of science begin their development by going through a 'prescience' phase, in which there is considerable disagreement among practitioners regarding the nature of the basic assumptions that underlie the field, and the sort of problems that should be investigated. Rival schools compete until the ideas of one school become generally accepted by other workers in the field and acquire the status of a paradigm. This is the beginning of the phase called 'normal science'. Scientists stop arguing about the fundamental principles, procedures and conventions that underlie their field and move on to more detailed, specialized work, using their newly acquired paradigm as a framework within which to operate. The great majority of scientists spend most if not all of their careers doing this type of work without ever questioning the validity of the paradigm on which it is based.

The career of a typical scientist is illustrative of normal science. From school to university, the scientist learns the basic facts, techniques, conventions and standards that constitute the paradigm on which his or her field is founded. The paradigm is studied in progressively greater detail and, if the scientist jumps over all the hurdles and enters postgraduate work, she or he tackles a demanding paradigm-based research project under the supervision of a fully qualified scientist. The successful completion of the research project endows the young scientist with a doctorate, which Kuhn sees as a licence to practise normal science by 'puzzle solving'-- tackling a succession of paradigm-based research projects.

According to Kuhn, paradigms are tenaciously held on to,

The Roots and Nature of Western Science

even when scientists are confronted with unquestionable evidence that signifies the paradigm may be starting to break down. Nevertheless, paradigm changes do sometimes occur, and Kuhn believes them to be an inevitable consequence of the nature of normal science. At first, the scientists who come across anomalous results invariably try to accommodate them within the current paradigm, either by modifying its detailed structure or by extending its scope by means of additional ad hoc hypotheses. There inevitably comes a time, however, when research generated by a particular paradigm starts to yield results that cannot be explained in terms of its basic concepts. The accumulation of anomalous results creates serious difficulties in attempts to reconcile the new discoveries with the existing paradigm without distorting it out of all recognition.

As normal science breaks down we enter the period of 'extraordinary science', where fundamental concepts are questioned and re-examined. This is the Kuhnian scientific revolution which follows the crisis caused by irreconcilable anomalies. From the competing paradigms that prevail, one will prove to be sufficiently superior to its rivals and consequently adopted as the new paradigm for the field. When this happens, a further period of normal science based on the new paradigm begins.

Kuhn gives copious examples from the history of science to support his thesis. For instance, replacement of the Earth-centred model of the solar system developed by the Greeks (the so-called Ptolemaic system) by the heliocentric system, having the sun at its centre. For almost two thousand years the Ptolemaic paradigm was used with great success to explain the movements of the various heavenly bodies which moved round the Earth in complicated orbits of cycles and epicycles. Anomalous results were accommodated within the paradigm by means of additional ad hoc hypotheses in the form of more and more epicycles. Eventually, however, the system became so cumbersome and unsatisfactory that some astronomers began to cast doubt on its validity.

Nicolaus Copernicus, a Polish cleric, gave clear expression to his doubts by publishing his extremely influential book, *De Revolutionibus Orbium Coelestium*, in 1543, in which the Sun replaced the Earth as the centre of our planetary system. As the validity of one of the oldest paradigms of science was directly challenged, a clash with the Church and the astronomical establishment was inevitable. In fact, one of Copernicus's followers, Giordano Bruno, was burned at the stake in 1600, while Galileo was threatened with a similar fate if he did not recant his views. It was only after a century or so of hard work by men such as Tycho Brahe, Kepler, Galileo and Newton that the heliocentric paradigm was accepted by the majority of astronomers.

The Darwinian revolution provides another example of paradigm shift in the Kuhnian sense. The upheaval in biological thinking was started by the English naturalist Charles Darwin in the middle of the nineteenth century. The contemporary paradigm of biology was the idea that living species were permanent and immutable, having been created in their present forms by God himself. The task of biologists was limited to taxonomic descriptions and the gathering of anatomical data.

The development of the science of geology, with the concomitant discovery of different types of fossils, began to cast doubts over the validity of the paradigm, doubts that became progressively more acute during the 1840s and 1850s as the implications of Darwin's discoveries came to be appreciated. The crisis point was the publication in 1858 of Darwin's book, *On the Origin of Species by Means of Natural Selection*, which led Darwin and the evolutionists to face prolonged and bitter opposition from both their fellow biologists and the Church. As in the case of the heliocentric theory, the evolutionary paradigm took a long time to gain general acceptance. In fact, most of Darwin's contemporaries

The Roots and Nature of Western Science

were never able to accept it.

Perhaps the most significant thing that Kuhn did was to bring the concept of values into science. The scientific community will sanction certain values that guide the choices of individual scientists when choosing between rival paradigms. The community values he outlines include accuracy, scope, simplicity and fruitfulness. He rightly stresses that the original protagonists of the new paradigm must have faith in their research programme, a word which many philosophers of science are loath to mention. None of his critics has been able to show that subjective factors do not play an important part in the evaluation of competing paradigms.

Kuhn's position, however, is very much a relativist one in that he claims that 'there is no standard higher than the assent of the relevant community'(p. 94). There is no place for absolute values in his depiction of science. Also, like Popper, he assumes that science is superior to other fields of inquiry: the paradigm of rationality. Indeed, he suggests that if a theory of rationality should clash with science then it is not science but our theory of rationality (which, of course, includes religion) that is at fault and needs to be changed.

> To suppose, instead, that we possess criteria of rationality which are independent of our understanding of the essentials of the scientific progress is to open the door to cloud-cuckoo land.
> Kuhn, in Lakatos and Musgrave, p. 264

For Kuhn, criticism of fundamental assumptions has no rational role to play in normal science, a discipline only becoming authentic mature science when religious or philosophical discussion of fundamentals is abandoned. In this way, Kuhn provides a rationale for scientists to pursue specialized puzzle-solving dissociated from all concern for philosophical and religious issues. That science transcends all other systems--spiritual and/or temporal--is the message we receive from Kuhn (as we did from Popper), but such a belief has now come under attack from contemporary philosophers of science.

Critique of Modern Science

Social and Political Dimensions

Although in modern times science is still highly esteemed, a significant number of writers have come to see modern science as dehumanizing, involving the inappropriate treatment of people and societies, as well as Nature. Many perceive the alleged neutrality and value freedom of science as a sham, and implicate modern science in the destruction of our environment resulting from technological advances. A runaway production-oriented technology leading to depletion of natural resources, an ever-increasing output of agricultural, industrial and marine pollution, not to mention the stockpiling of formidable arsenals of nuclear, biological and chemical weapons, are seen as indicators of an impersonal threat to mankind's future. Modern science, with its roots in capitalist ideology, is seen as the central force which has unleashed such global problems.

Paul Feyerabend, one of the more widely read philosophers of science, has consistently opposed and derided Popperian and Kuhnian venerations of science. He views science as nothing more than an ideology playing a role akin to that which Christianity played in Western society a few hundred years ago and from which we need to be liberated. There is a separation between state and Church, but no separation between state and science. Therefore, writes Feyerabend, we need to 'free society from the strangling hold of an ideologically petrified science just as our ancestors freed us from the strangling hold of the One True Religion' (Feyerabend, *Against Method*).

Science, in other words, should not be given preference

over other forms of knowledge or traditions. Modern science does not possess features that render it distinct from and superior to voodoo or astrology. In fact, Feyerabend finds it curious that while an American can now choose the religion he likes, he is still not permitted to demand that his children learn magic rather than science at school.

Although Feyerabend's anarchistic theory of knowledge leads him to the absurd belief that a state should be ideologically neutral, he nevertheless convincingly argues against the idea of a universal scientific method. Science cannot provide rules adequate for guiding the activities of scientists, for 'the idea that science can, and should, be run according to fixed and universal rules is both unrealistic and pernicious'.

Nicholas Maxwell (1984) is another not-so-well-known philosopher of science who has approached the problem of modern science from a different perspective. He labels the philosophical base of modern science as the philosophy of knowledge, espousal of which has resulted in a major intellectual disaster at the heart of Western science, technology, scholarship and education. This long-standing intellectual disaster (ever since the time of Bacon) has much to do with the human disasters of our age, our incapacity to tackle more humanely and successfully our present worldwide problems. Maxwell therefore advocates 'the need to put into practice a profound and comprehensive intellectual revolution affecting to a greater or lesser extent all branches of science, technology, scholarship and education' (*From Knowledge to Wisdom*, p. 3).

The central claim of Maxwell's thesis is that modern scientific inquiry is profoundly and damagingly irrational because it gives intellectual priority to the task of improving knowledge. In fact, priority should really be given to intellectually more fundamental problems of living, problems of knowledge and understanding tackled as rationally subordinate.

Maxwell cites the twentieth-century mismatch between

progress in scientific knowledge and progress in cultivating human brotherhood. The actual institutional structure of academic inquiry, therefore, needs to be changed, from knowledge to wisdom. Indeed, the philosophy of wisdom is his intellectual antidote to the current prevalence of the philosophy of knowledge. The philosophy of wisdom stresses the intellectually fundamental character of articulating problems of living, proposing and criticizing possible solutions, rather than giving precedence to secondary problems of knowledge and understanding of the natural world. To quote Maxwell at length:

> During the twentieth century mankind has made extraordinary progress in scientific knowledge, and in technological and industrial development. During the same period, mankind has committed horrifying crimes against itself, in that millions upon millions of people have suffered and died as a result of war, tyranny, concentration camps, mass executions, economic exploitation and increasingly unjust distribution of the world's resources. A major reason for this glaring discrepancy between what has been achieved in knowledge and in life is that during the last two or three centuries--and especially during the twentieth century in the developed world--mankind has succeeded only in developing socially influential organized inquiry in accordance with the philosophy of knowledge and has thus failed to develop organized inquiry in accordance with the philosophy of wisdom. As a result, specialized knowledge has flourished, but social wisdom in the world has faltered. If we are to progress towards a wiser world it is essential that science, technology, scholarship and education in schools, universities and research establishments throughout the world be transformed to accord with the edicts of the philosophy of wisdom. If organized inquiry is developed in this way, then we may reasonably hope to make gradual progress towards a more just, humane, co-operative--and even loving--world.
>
> *From Knowledge to Wisdom*, p. 152

Other writers have gone further and argued that science is an industrialized enterprise, generating oppression and inhumanitarian technologies. Rose and Rose, for example, have claimed that there is ideology in and of science, and that the technologies generated by the enterprise are part and parcel of the total system of Western science. The supposed neutrality of scientific methodology, assumed objectivity and freedom from values is a grotesque misconception at best and an outright falsehood at worse. In fact, Mitroff showed that the putative objective criteria of science are far outweighed by subjective elements which constantly come into play both in the construction and evaluation of scientific theories. Critics inside and outside the scientific establishment have, in other words, consistently argued that scientific knowledge, and ignorance, are socially constructed. Scientific theories are spawned in a specific social and cultural milieu, politics often dictating the way in which scientific discoveries will be used.

A clear example of such a context is provided by Gideon Freudenthal who ably constructs a social explanation of some aspects of Newton's physics. He does not seek to give a social derivation of the entire content of the *Principia Mathematica* (Newton's seminal work published in 1687), but demonstrates how certain significant assumptions at work in the book have their origins in social relations. For example, in propounding the concept of an absolute space, Newton assumes that the material world is composed of equal particles, each possessing the same essential properties-- properties that a particle would continue to possess even if it were alone in empty space. Freudenthal traces Newton's assumption back to the individualist conceptions of society that emerged in the seventeenth century as feudal society gave way to early forms of capitalist society. The market and the individual played a central role in this society. Individuals were believed to possess essential properties, properties which they were assumed to possess independently of their existence in society. Freudenthal shows that the general philosophical principle which underlay certain fundamental aspects of Newtonian physics and the emerging capitalist society was the same, having the same social origin: that the properties of wholes are to be explained in terms of the essential properties of their parts.

In reference to the tenacious belief in scientific objectivity, American sociologist Theodore Roszak talks of 'the myth of objective consciousness'. According to this myth, there is only one way of gaining access to reality: to cultivate a state of mind cleansed of all subjective distortion and all personal involvement. Roszak believes that only what flows from this consciousness qualifies as knowledge. As if to complement Roszak's view, the philosopher of science Michael Polanyi wrote that 'objectivism has totally falsified our conceptions of truth'.

Critique from the Marxist Perspective

In the 1960s, some Cambridge scientists helped to found the British Society for Social Responsibility in Science in order to popularize their views on science and socialism. This was at a time when many scientists in Britain, North America, France, West Germany, Belgium and Italy were increasingly concerned about their findings being used for destructive purposes. Their fears were unhappily confirmed by the results of the war in Vietnam and in the proliferation of nuclear weapons. The human destruction and environmental devastation caused by the use of chemical and biological weapons in Vietnam was too much to countenance.

There were two main kinds of responses to this scenario. Many scientists argued that science was being manipulated by politicians to achieve sinister political objectives. It was the moral responsibility of the scientific establishment, therefore, to ensure through collective agitation that the results of their research were not applied in such a way as to cause human destruction and environmental devastation. The other response was to see science itself as the problem, rather than an external political hand manipulating it. These scientists argued that science is erected on ideological foundations, that scientific theories help legitimate the status quo, that of the capitalist political economy. That there is ideology within and of science is essentially the Marxist view of science.

Hilary and Stephen Rose have argued, for example, that many neurobiological theories and associated technologies, such as drug therapy and IQ testing, are fundamentally biologistic. Biologism, through reducing all aspects of the human condition to the level of biological principles is parochial in its perspective. More fundamentally, however, by providing a biological explanation for aggression, war, love and hate, it implies that it is absurd to change ourselves or the world. For all its claims to be scientific, argue the Roses, biologism is ideological in the sense that it helps to legitimate the status quo.

The Roses have gone further than this and argued that modern science is sexist and racist. From reproductive technology to genetic engineering and hormone time capsules, sexism is ingrained in current scientific developments. In fact, Brian Easler, another well-known Marxist philosopher of science, puts forward the thesis that nineteenth-century physicians promulgated the belief that women are devoid of passion. This idea was legitimized and reinforced by subsequent medical ideas and practice.

That modern science is based on capitalist ideology which is racist and sexist is agreed by all Marxist historians of science. Some feel that within this ideological framework some science is still, or conceivably can be, objective. Others, however, reject this view and claim that all science is ideological, that there can be no objectivity to science. Such a proposition implies that science does not provide a view of the world which corresponds to reality but merely manifestations of social relations. The Roses, among others, are critical of such an extreme view, arguing that science is relatively autonomous and analysis of the properties of physical bodies is possible despite ideological intrusions into the bodywork of science. By saying this they claim to be restating Marx's position: that science is not neutral but is able to provide us with a knowledge of reality. Moreover, because science is continually transforming conceptually, it cannot simply be reduced to an ideology that reflects what happens in class struggles.

The discussion of the Marxist perspective of science would not be complete without mentioning the views of the late J.D. Bernal who was a committed communist. As a leading representative of the humanistic Marxist vision of science, Bernal's central argument is that science is inherently progressive but can reach its full potential benefit only in a communist society. Technological transformation was to be achieved through a rationally planned and organized science, something which only a trained scientific elite could ensure. For his deeply influential views, Bernal was awarded the Stalin Peace Prize in 1953. He subsequently joined a group of other like-minded socialist scientists who played a significant role in changing the British Labour Party's perpsective on science and technology. All the big names in Labour party politics--Hugh Gaitskell, Harold Wilson, James Callaghan and Richard Crossman--used to attend these meetings, which helped them to formulate a science policy that modernized Labour's image in preparation for the 1964 general election. It was Bernal's socialistic vision which was the basis of Wilson's immortal phrase 'white hot technological revolution'.

The quintessence of Bernal's vision is epitomized in the last paragraph of his famous book:

Already we have in the practice of science the prototype for all

The Roots and Nature of Western Science

human common action. The task which the scientists have undertaken--the understanding and control of nature and of man himself--is merely the conscious expression of the task of human society. The methods by which this task is attempted, however imperfectly they are realized, are the methods by which humanity is most likely to secure its own future. In its endeavour, science is communism. In science men have learned consciously to subordinate themselves to a common purpose without losing the individuality of their achievements. Each one knows that his work depends on that of his predecessors and colleagues and that it can only reach its fruition through the work of his successors. In science men collaborate not because they are forced to by superior authority or because they blindly follow some chosen leader, but because they realize that only in this willing collaboration can each man find his goal. Not orders, but advice, determines action. Each man knows that only by advice, honestly and disinterestedly given, can his work succeed, because such advice expresses as near as may be the inexorable logic of the material world, stubborn fact. Facts cannot be forced to our desires, and freedom comes by admitting this necessity and not by pretending to ignore it. These are things that have been learned painfully and incompletely in the pursuit of science. Only in the wider tasks of humanity will their uses be found.

Bernal thus equates science with communism. Although this is a radical notion, there are also radical non-Marxist perspectives, such as that of J.R. Ravetz. Ravetz is a contemporary philosopher/sociologist of science who has significantly contributed to the radical transformation of our understanding of science over the past few decades. He believes that the conventional picture of science as the sole neutral way of studying nature and the objective pursuit of truth is erroneous and therefore untenable. Science is a socially conditioned activity, with values, belief systems, social conditions, professional interests, peer group pressure, secrecy, property rights and naked ego all playing their part in the social construction of science, and ignorance. Ravetz maintains that 'the commitment to humanity must be at the core of the scientific endeavour, if it is to be worth pursuing at all' (*Merger of Knowledge*, p. 3).

In propounding his thesis, Ravetz admits that the ultimate motive of the strategic planning of scientific activity in close alignment with the ruling political/economic structures of society may well be the improvement of the condition of mankind. The point, however, is that the science done, or perhaps more significantly the science *not* done, reflects the values of a society as they are realized in its dominant institutions. Taking up the latter point, Ravetz argues that public ignorance concerning important scientific issues of ethical significance is not due to their intellectual impenetrability. On the contrary, public ignorance is socially constructed by decisions taken in leading institutions of state and of science to neglect certain problems in favour of others.

> Such problems will usually *not* be those promising prestige and rewards to a scientific elite, but rather those involving diffuse, imperceptible, chronic or delayed effect of the unintended by-products of the industrial system. In that sense, our scientific-technological establishment moulds public awareness, by negative means, as surely as did the theological establishments of earlier times by indoctrination and prohibitions. The social construction of ignorance is a phenomenon of our modern period, all the more important because it happens unnoticed and in contradiction to the received ideology of science as the bearer of truth.
>
> Ravetz, *Merger of Knowledge*, p. 225

One thing that should be fairly evident from these critiques is that the connection between Western culture and Western science is a deep-rooted one. I have also shown this to be the case in 'Ismail Al-Faruqi and Ziauddin Sardar: Islamization of Knowledge or the Social Construction of New Disciplines?' (*MAAS Journal of Islamic Science*, June 1989), where I trace the social origins of phrenology in early nineteenth-century Edinburgh and the development of modern medicine in France. I show how the origins of phrenology as a pseudo-intellectual discipline had a powerful social and cultural context which led to conflict between incommensurable worldviews, involving a wide range of theological, philosophical, scientific and methodological issues.

The social and cultural bases of some of the far-reaching innovations in the structure of medical knowledge in Paris in the early part of the nineteenth century are also explored in this paper. All this is done within the context of showing that the late Ismail Al-Faruqi's initially unquestionable acceptance of Western scientific disciplines was an intellectual mistake. Disciplines are not God-given epistemological categories, but are created, developed and evolved within a social and cultural milieu, that is, within the matrix of a particular worldview.

The worldview of modern science is reductionistic, ruthlessly atomizing all systems in order to study their basic functions. The holistic perspective is marginalized in this process, and reason is given supreme importance in its reductive methodology.

The Muslim critique of modern science, therefore, centres on one basic conception: that modern, Western science incorporates a fundamentalist attitude to reason. The whole enterprise has become reduced to a tool of reduction, operating within the confines of a very narrow epistemological system. This necessarily means that only those aspects of a phenomenon which are amenable to pure reason are really worthy of investigation. Crucial issues of ethics and morality are marginalized so as not to 'pollute' the neutrality of the enterprise. Reason has become the only way of knowing and the only arbiter of truth, whether considered absolute or approximate being immaterial. By raising reason to the level of a god, modern science is implicitly centralizing its position as a way of knowing which transcends all else, including divine revelation. The edifice of the whole enterprise is built upon this exclusivist and reductive use of reason. It is in this sense that modern science has assumed a fundamentalist position, one which Ziaddin Sardar says 'can be defended only by declaring war on everything and everyone else' (*Information and Muslim World*, p. 2).

It should be apparent that the critics of modern science--an enterprise developed and moulded by the Western world over the last four hundred years--claim, in a variety of contexts, that a more socially equitable use of the scientific knowledge that we have is a more pressing problem than the production of more scientific knowledge. They believe that the economic and military interests of government agencies and the allied interests of multinational corporations constitute a major driving force underlying the direction of development of Western science. They wish to see science develop in directions more in keeping with the needs and interests of ordinary people, and evaluated and articulated with reference to other interests and values. In essence, the need to grasp the limitations and scope of scientific knowledge is stressed in their writings. Islamic science does all of this and more.

Various Positions on Science

1. *Infallible empiricism.* Infallible theoretical knowledge can be arrived at from empirical data alone, using inductive methods.

2. *Fallible empiricism.* Sound knowledge can be arrived at by means of induction from empirical data. Admittedly, this knowledge may be fallible, requiring subsequent revision.

3. *Aprioristic empiricism.* Almost infallible scientific knowledge

can be obtained by empirical investigation combined with deducing basic metaphysical principles from pure reason.

4. Infallible (naive) inductivism. Law and theories, once formulated, can be firmly established as true by inductive derivation from empirical data.

5. Fallible inductivism. Laws and theories, though derived and verified inductively, nevertheless remain fallible or probabilistic, open to revision.

6. Hypothetico-deductivism. There is no such thing as inductive rules of reasoning from data to theories. Rather, hypothetical laws and theories are to be assessed by means of the empirical verification and falsification of propositions deduced from them.

7. Falsification. There is no verification, or even tentative verification, in science. All scientific knowledge is conjectural in character, progress being made only through the empirical falsification of theories.

8. Theoretical pluralism. Science is a worldview which happened to replace Christianity (or religion in general), all worldviews having equal validity.

9. Paradigmism. Initially empirically successful theories are accepted and developed within a research tradition until outmanoeuvred by an empirically more progressive paradigm.

10. Implicit intellectualism. The empirical assessment of scientific results cannot be adequately encapsulated in any neat set of explicit rules or methods.

11. Metaphysical relativism. The central task of scientific inquiry is to devote reason to the enhancement of wisdom--wisdom being understood as the desire, the active endeavour, and the capacity to discover and achieve what is desirable and of value in life, both for oneself and for others. To this end, the implicit metaphysical assumptions of science, which may change over time, must be made explicit.

12. *Metaphysical determinism (Marxist perspective).* There is an ideology in and of science, science being a socially-conditioned activity. The conceptual tools with which we seek to understand scientific phenomena are determined by the socio-economic imperatives that operate in society.

13. *Islamic science (holistic, based on divine revelation).* The scientific enterprise is not erected upon certain relative metaphysical assumptions, but upon a foundation of conceptual certainties. *Tawheed* (unity of God) is the macroparadigm, and *khilafa* (man's trusteeship) the concomitant principle within the framework of which science is pursued for promoting *adl* (all forms of justice) in the public interest (*istislah*).

2

Islamic Science

The New Paradigm

Why Islamic Science?

> If science is not the unique intellectual construct, which until so recently it was portrayed, if the history of science is not the history of iterative movements towards the truth about the natural world, but rather the history of various social constructions of reality mediated through science, scientists and society, then there exists the possibility of an Islamic science that will be one, or more likely a series of facets of a multidimensional world of nature, all of which are imbued with the very essence of Islamic society.
>
> Glyn Ford, in Sardar, *Touch of Midas, pp.* 34-5

If science is indeed value-laden with a significant subjective (as well as objective) component, then it can be developed with a specific cultural style and emphasis. That is, in an Islamic society the values shaping scientific and technological endeavour would have to be Islamic values, which in a nutshell is what we mean by the concept of Islamic science.

It is important to understand in this connection that there is no dichotomy or conflict between Islam and science: the well-documented hostility between institutionalized Christianity and the scientific enterprise has no parallel in Islamic history. Faith and rationality find their perfect fusion in Islam. Science and technology, economy and politics, are encompassed within its worldview, Islamic ethics and values constituting the strand of unification that permeates all human activity. It is, in short, an holistic system that touches every aspect of human endeavour.

The concept of an Islamic science has not dawned upon many Muslim academics and intellectual circles as yet. Some do not see the relevance of it because their minds are steeped in the belief in a neutral and value-free universal enterprise that transcends social and cultural differences. Although unable to articulate it, let alone identify it, they have faith in the existence of a universal, ahistorical scientific method, so the question of 'Western' or 'Occidental' or 'Islamic' science becomes immaterial, or at best a matter for insignificant philosophical musings.

There are other Muslim scientists, however, who menacingly seek to legitimize modern science by equating it with the Quran. They do this by pointing out that the Quran puts great emphasis on the pursuit of knowledge about nature, even producing statistical evidence in corroboration. When the factual statements mentioned in the Quran pertaining to natural phenomena find their confirmation in modern science, euphoric feelings are experienced. The implication is that the veracity and authenticity of the Quran is based upon the empirical criteria of modern science which are therefore assumed to have universal and eternal validity.

Such an intellectually misguided attitude is most prominent in the works of the French surgeon, Maurice Bucaille, who quite articulately expounds his thesis in his widely read and admired book, *The Bible, the Quran and Science*. In it he undertakes a detailed analysis of the holy scriptures in the light of modern knowledge and comes to the conclusion that

> The Quran most definitely did not contain a single proposition

Islamic Science: The New Paradigm

at variance with the most firmly established modern knowledge, nor did it contain any of the ideas current at the time of the subjects it describes. Furthermore, however, a large number of facts mentioned in the Quran were not discovered until modern times. So many, in fact, that on November 9, 1976, the present author was able to read before the French Academy of Medicine a paper on the 'Physiological and Embryological data in the Quran'. The data, like many others on different subjects, constitutes a veritable challenge to human explanation--in view of what we know about the history of the various sciences through the ages. Modern man's findings concerning the absence of scientific error are therefore in complete agreement with the 'Muslim exegetes' conception of the Quran as a Book of Revelation. It is a consideration which implies that God could not express an erroneous idea.

Bucaille, p. 7

The inherent damage in this type of thinking should be quite apparent. Whether or not an idea is correct or erroneous depends upon whether or not it is compatible with modern scientific findings. The rules of the game are therefore laid by science at every stage, with the inevitable implication that failure (incompatibility) is possible. If the Quran at present is in complete accord with modern findings, as claimed by Bucaille, then it must always be in a state of non-conflict if it is to maintain its authenticity with respect to the Bucaillean criterion. Apart from raising the status of science to that of supreme intellectual arbiter, the argument itself is a naive inductivist one, resting on a precarious foundation and lacking intellectual rigour. For a Muslim, the Quran needs no justification from modern science; it is an eternally valid, universal source of guidance.

There are yet other Muslim scientists who, while maintaining a belief in the universality, neutrality and value freedom of modern science, proclaim that its functions can be modified to serve Islamic ideals and Muslim societies. In other words, they do not see science as the intellectual arm of capitalist ideology. With the sincerity of naivete, they wish to take Western science as it is and transplant it into Islamic culture. After eliminating any amoral, mechanistic and hedonistic tendencies, such an intellectual transplant within the parameters of Islamic consciousness is expected to lead to a renaissance of science in the Muslim umma.

Taken at face-value, the idea of 'science in an Islamic context' is deceptively appealing. It makes science subservient to the goals and values of Muslim societies, operating within the parameters of Islamic ethics. There is, however, a fundamental flaw lying at the very heart of the whole idea. It is this. Modern science, as it exists in the West, developed hand in hand with capitalist ideology. Separation of religion and state has been a central tenet of this ideology, having led to the inevitable bifurcation and subsequent confrontational postures of religion and science. The well documented hostility between Western science and Christianity is, therefore, no historical accident. It was the values of capitalist ideology, of which modern science is a product, that really clashed with Christian dogma. The clergy were unable to perceive this. What they saw openly manifested were the 'unacceptable' findings of modern science, and not the ideology which actually conceived it.

What this means is that the entire system of science that exists today is deeply entrenched in Western values and culture and in turn derived from capitalist ideology. Writers such as Ravetz and Mitroff have even argued that the experimental and quantitative techniques of modern science cannot escape the onslaught of these values. Far from being an asset to the Muslim umma, the concept of 'science in an Islamic context' would actually be detrimental to the progress of Muslim societies.

We therefore need a science that is conceived from within the worldview of Islam, whose processes and methodologies incorporate the spirit of Islamic values, and which is therefore performed solely for the pleasure of Allah. Such a science would serve the needs and requirements of Muslim societies and operate within the frame work of Islamic ethics. The nature and style of this science would be radically different from that of modern science. Because it is a product of Islamic civilization, such an enterprise is appropriately designated an Islamic science.

Expounding Islamic Science

Islamic science is still very much in the developmental stage and in the process of being eloquently articulated. This point was most starkly illustrated by the Muslim biologist Munawar Ahmad Anees who found it decidedly easier to list those things which Islamic science is not rather than expounding what it actually is. He wrote that Islamic science is not:

1. Islamized science, for its epistomology and methodology are the products of Islamic worldview that is irreducible to the parochial Western worldview.
2. Reductive, because the absolute macroparadigm of tawheed links all knowledge in an organic unity.
3. Anachronistic, because it is equipped with future-consciousness that is mediated through means and ends of science.
4. Methodologically dominant, since it allows an absolute free-flowering of method within the universal norms of Islam.
5. Fragmented, for it promotes polymathy in contrast with narrow disciplinary specializations.
6. Unjust, because its epistemology and methodology stand for distributive justice with an exacting societal context.
7. Parochial, because the immutable values of Islamic science are the mirror images of the values of Islam.
8. Socially irrelevant, for it is 'subjectively objective' in thrashing out the social context of scientific work.

9. Bucaillism, since it is a logical fallacy.
10. Cultish, for it does not make an epistemic endorsement of Occult, Astrology, mysticism and the like.

> Anees, 'What Islamic Science is Not', p. 19-20

We can, therefore, confidently say that the absolute macroparadigm, as Anees calls it, for Islamic science is Islam. Indeed, it is erroneous to view Islamic Science as a subspecies of the scientific macroparadigm. The Islamic worldview provides the intellectual matrix within which scientific ideas, thoughts and theories are conceived, developed and elaborated. Any science conceived and developed within a worldview that rejects the universal norms of Islam is, then, by definition epistemologically alien to Islamic science.

Knowledge is a whole entity under Islamic science, pursued in its entirety using an integrative approach. Such a holistic view is derived from the all-encompassing, all-pervasive concept of tawheed, which was translated and internalized in the pursuit of knowledge by early Muslim scholars. Thus, it was not uncommon for a Muslim scholar to be an astronomer, a medical practitioner and a philosopher at the same time. Science was practised to find out more about Allah's creation and to earn His pleasure, rather than for reasons of power, monopolization, money and exploitation.

Islamic science is therefore universal and not parochial. Due to its nature and style, it has successfully avoided conflicts of revelation and reason, all too common throughout the history of Western ecclesiastical antagonism.

This avoidance of conflict is not through the misguided efforts of scientifically illiterate apologetic Muslims who consistently employ scientific data to uphold the truth of revelation. In doing so they grant revelation an inferior position and Western science a superior one. Rather, it is the result of not tempering the revealed message with the transient and ever-changing discourses of modern science.

The values of Islamic science are immutable because they

Islamic Science: The New Paradigm

are derived from the universal worldview of Islam. It is ironic that Western science, in spite of its vociferous claims of being value-free, neutral and objective, is in fact based upon transient, ever-chainging values of the establishment, as expounded by Kuhn in his brilliant *Structure of Scientific Revolutions*. In the case of Islamic science, however, value clarification is not a problem. There are no implicit and unstated assumptions. It operates in a universal value system derived from the macro-paradigm of the Islamic worldview.

In an attempt to define the immutable values on which Islamic Science is based, a seminar on 'Knowledge and Values' was held under the auspices of the International Federation of Institutes of Advance Study (IFIAS) in Stockholm during September 1981. The participants isolated ten Islamic concepts which together form the value framework of Islamic Science:

1. tawheed (unity, or oneness of Allah)
2. khilafa (the trusteeship of man)
3. ibadah (worship)
4. ilm (knowledge)
5. halal (permissible or praiseworthy)
6. haram (forbidden or blameworthy)
7. adl (justice)
8. zulm (tyranny)
9. istislah (public interest)
10. dhiya (waste)

Islamic scientific enterprise would progress by taking the positive values as its guiding principles and deciding its research priorities as well as project implementation on their basis. The function of the negative values of haram, zulm and dhiya is effectively to maintain all scientific activity within ethically acceptable research parameters. If the legitimate boundaries of Islamic science are ever in danger of being violated, then these negative values are operationalized to maintain the ethical homeostasis of Muslim society.

The central paradigmatic concepts of Islamic science are tawheed, khilafa and ibadah, concepts which explain the role and purpose of human life, making sense of men and women, life and the universe. Muslim scientists and Islamic scientific centres and institutions would have as their central objective the promotion of adl and istislah, while at the same time discouraging and undermining zulm and dhiya. Every research programme put forward as a discussion document for possible implementation would be thoroughly assessed to ensure that it is not economically, socially or culturally unjust. All destructive projects and research endeavours (physically, socially, economically, culturally, spiritually and environmentally) would not be permitted, because such science and technology promotes rampant consumerism with the concomitant accumulation of wealth in fewer and fewer hands. We witness today the tangible effects of such a runaway technology in the alienation and dehumanization of a major segment of humanity. This is a major characteristic of zalim (tyrannical) science, resulting in the destruction of human, environmental and spiritual resources, thereby generating waste (dhiya). As Sardar has aptly said:

> While this rather theoretical model of Islamic science needs much further work, it is clear that it can form the basis of a practical science policy for Muslim countries. Islamic concepts, as the early history of Islam demonstrates so brilliantly, do not only have analytical value: they are also intrinsically pragmatic. Without operationalizing these key concepts, it is difficult for a society, or a civilization, to claim that it is Islamic. Thus the model of Islamic science developed at the Stockholm seminar has a strong practical value. Apart from shaping science policies of Muslim countries, it can also be used as a criteria for examining the nature and content of Western science and determining the value of its various components for Islamic societies. More generally, it can be used as the framework of a

Islamic Science: The New Paradigm

critique of modern science--a critique that would highlight the fact that the inhuman rationality of modern science can be tamed, indeed synthesized, with a humane vision of knowledge to the benefit of all mankind.

Sardar, *Islamic Futures*, p. 176-7

Islam is a total system, a religion, a culture and a civilization all at once. It is an holistic system which touches every aspect of human endeavour. Its ethics and values permeate all human activity, including science, as we have just seen.

It has already established that there is no conflict between Islam and science when by 'science' we mean a rational and empirical method for studying natural phenomena. The only time when conflict arises is when science and its methodology are made into an all-embracing transcendant value at the expense of Islamic values. The pursuit of knowledge in Islam is only a means of acquiring a deeper understanding of the Creator, and solving the problems of the Muslim community. It is not an end in itself. Science is therefore not pursued for science's sake, but for seeking the pleasure of God through trying to understand His signs, and by extension Him.

The Quran stresses that man is an integral part of the natural world and has been endowed with the ability, to a large but limited extent, to control the forces of nature. Nature, however, is not there to be exploited for the unethical desire for domination, but for utilizing its resources in the nobler interest of serving humanity. Hence, the Quran places knowledge about the natural world into its proper context; that is, within a framework of total human experience. Reason and the pursuit of knowledge have an important place in an Islamic society but are subservient to quranic values and ethics.

This is not the context, unfortunately, of modern science, where revelation and reason are divorced. Reason is often considered to be supreme, with the inevitable relativization of

*A diagrammatic representation of the Islamic concepts
which embrace and describe
the nature of scientific enquiry*

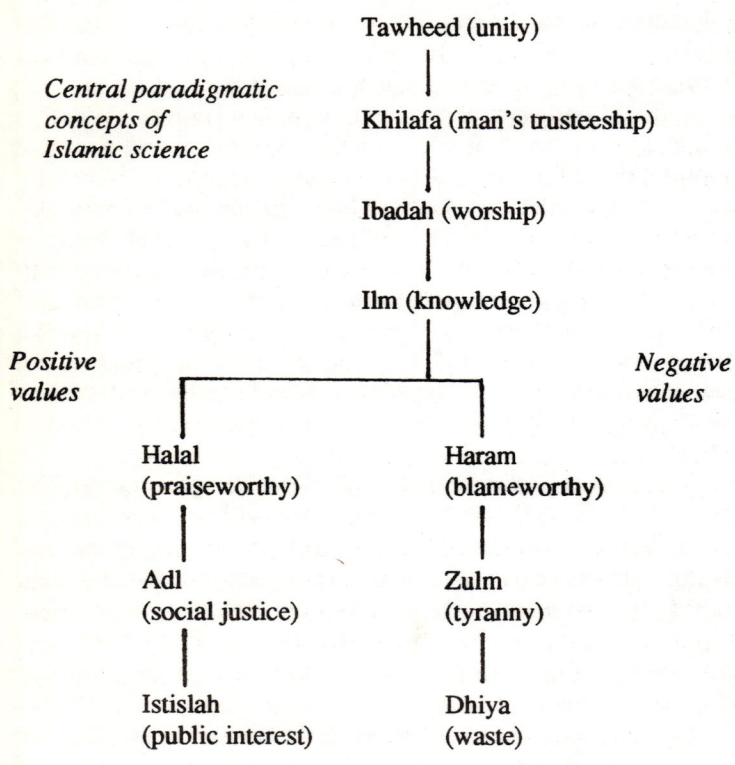

ethics and values. The apostles of modern science reject or ignore fundamental religious concepts such as man's trusteeship and the Day of Judgement, which are central to the principle of human accountability for all worldly activities.

Belief in a Creator makes Muslim scientists more conscious of their activities. It places their reason under the authority of God to whom they are responsible for all their actions. In Islam, therefore, there is no divergence between the means and ends of science. Both are subject to the ethical and value parameters of the faith. It necessarily follows that as scientists are accountable to God for their activities, they must effectively discharge the social function of science in service to the community, and at the same time protect and promote its ethical and moral institutions. Thus the Islamic approach to science is built on an absolute moral and ethical base, upon which a dynamic edifice is erected. Reason and objectivity are promoted in the pursuit of scientific knowledge at the same time as placing the intellectual endeavour within the boundaries of Islamic ethics and values.

Promotion of such eternal Islamic values as khilafa (man's trusteeship), ibadah (worship) and adl (justice), is the subjective aspect of Islamic science. Human emotions, biases and prejudices are laid aside to make way for the pursuance of such sublime goals through scientific inquiry. The objectivity of the enterprise comes into play through the research methods and procedures used which encourage the free formulation, testing and analysis of hypotheses, with the subsequent modification and retesting of theories if necessary.

Because science describes and is successful in explaining a very limited aspect of reality, it serves powerfully to remind us of the weaknesses and limitations of human capacity. The Quran also constantly reminds us to be cognizant of our limitations before getting carried away by the successes of our scientific discoveries and the astounding results of scientific research:

> Does not man see that it is We who created him from sperm? Yet behold! he (stands forth) as an open adversary! And he makes comparisons for Us, and forgets his own (Origin and) Creation: He says, 'Who can give life to (dry) bones and decomposed ones (at that)?' Say, 'He will give them life Who created them for the first time! For He is well-versed in every kind of creation! The same Who produces for you fire out of the green tree, when behold! you kindle therewith (your own fires) Is not He Who created the heavens and the earth able to create the like thereof?--Yea, indeed! For He is the Creator Supreme, of skill and knowledge (infinite)! Verily, when He intends a thing, His Command is, 'Be', and it is! So glory to Him in Whose hands is the dominion of all things: and to Him will you be all brought back.
> Quran 36:77-83

A Comparison between Western Science and Islamic Science

Norms of Western Science	Norms of Islamic Science
1. Faith in rationality.	1. Faith in revelation.
2. Science for science's sake.	2. Science is a means for seeking the pleasure of Allah; it is a form of worship which has a spiritual and a social function.
3. One all-powerful method, the only way of knowing Reality.	3. Many methods based on reason as well as revelation, objective and subjective, all equally valid.
4. Emotional neutrality as the key condition for achieving rationality.	4. Emotional commitment is essential for a spiritually and socially uplifting scientific enterprise.

Islamic Science: The New Paradigm

5. Impartiality: a scientist must concern himself or herself only with the production of new knowledge and with the consequences of its use.

5. Partiality towards the Truth: that is, if science is a form of worship a scientist has to concern herself or himself as much with the consequences of their discoveries as with their production; worship is a moral act and its consequences must be morally good; to do any less is to make a scientist into an immoral agent.

6. Absence of bias: the validity of scientific statement depends only on the operations by which evidence for it was obtained, and not upon the person who makes it.

6. Presence of subjectivity: the direction of science is shaped by subjective criteria: the validity of a scientific statement depends both on the operation by which evidence for it was obtained and on the intent and worldview of the person who obtained it; the acknowledgement of subjective choices in the emphasis and direction of science forces scientists to appreciate their limitations.

7. Suspension of judgement: scientific statements are made only on the basis of conclusive evidence.	7. Exercise of judgement: scientific statements are always made in the face of inconclusive evidence; to be a scientist is to make expert, as well as moral judgement, on the face of inconclusive evidence; by the time conclusive evidence has been gathered it may be too late to do anything about the destructive consequences of one's activities.
8. Reductionism: the dominant way of achieving scientifiic progress.	8. Synthesis: the dominant way of achieving scientific progress; including the synthesis of science and values.
9. Fragmentation: science is too complex an activity and therefore has to be divided into disciplines, subdisciplines and sub-subdisciplines.	9. Holistic: science is too complex an activity to be divorced and isolated into smaller and smaller segments; it is a multidisciplinary, interdisciplinary and holistic enterprise.
10. Universalism: although science is universal, its primary fruits are for those who can afford to pay, hence secrecy is justified.	10. Universalism: the fruits of science are for the whole of humanity and knowledge and wisdom cannot be bartered or sold; secrecy is immoral.

Islamic Science: The New Paradigm

11. Individualism: which ensures that the scientist keeps his distance from social, political and ideological concerns.	11. Community orientation: the pursuit of science is a social obligation (fard kifaya); both the scientist and the community have rights and obligations on each other which ensure interdependence of both.
12. Neutrality: science is neutral, it is neither good nor bad.	12. Value orientation: science, like all human activity is value laden; it can be good or evil, 'blameworthy' or 'praiseworthy'; science of germ warfare is not neutral, but evil.
13. Group loyalty: production of new knowledge by research is the most important of all activities and is to be supported as such.	13. Loyalty to God and His creations: the production of new knowledge is a way of understanding the 'signs' of God and should lead to improving the lot of His creation--man, wildlife and the environment. It is God who has provided legitimacy for this endeavour and therefore it must be supported as a general activity and not as an elitist enterprise.
14. Absolute freedom: all restraint or control of scientific investigation is to be resisted.	14. Management of science is an invaluable resource and cannot be allowed to be wasted and go towards an evil direction; it must be carefully managed and planned for and it should be subjected to ethical and moral constraints.

15. Ends justify the means: because scientific investigations are inherently virtuous and important for the well-being of mankind, any and all means--including the use of live animals, human beings and foetuses--are justified in the quest for knowledge.	15. Ends do not justify the means: there is no distinction between the ends and means of science; both must be halal (permitted), that is, within the boundaries of ethics and morality.

(From Sardar, *Explorations in Islamic Science*, pp. 95-7)

Contemporary Issues: the Environment

Increased levels of atmospheric carbon dioxide exacerbating the global greenhouse effect and depletion of the stratospheric ozone layer which controls the amount of ultraviolet light entering the earth's atmosphere are problems that politicians, scientists, priests and journalists constantly discuss. The rapid deterioration of the human environment, one of the most striking manifestations of the crisis of Western science and technology, is, in fact, a crisis of values. To identifiy the underlying causes of our current environmental predicament, therefore, we must look to our attitudes which are ultimately shaped by our worldviews.

If the environmental crisis is indeed a crisis of values then the threat to our planet cannot be checked simply by introducing more strict legislation against pollution (the 'polluter pays' principle), industrial waste and nuclear spill. The adoption of national and international conservationist policies may help stem, to some extent, the degradation of our environment, but it will not root out the problem. The whole ecological imbalance is a reflection of the spiritual crisis of modern civilization itself. Hence, it requires a fundamental

Islamic Science: The New Paradigm

reassessment of our way of life, of our very conception of the relationship between humans, life and universe. It is my conviction that Islamic values and ethics provide a most effective and comprehensive answer to the absurdities of our environmental predicament.

With the concept of tawheed, the assertion of God's unity, the whole of the ecological crisis is brought under moral control. Nature and ethics are integrated and the unity of purpose and goal, intent and action, is achieved.

From tawheed emerge the concepts of khilafa and amana. The quranic concept of khilafa underpins the entire rationale of Islamic environmental ethics. Men and women have been given the amana, or trust, from God as His vicegerents, and are therefore responsible for looking after the vast stocks of our planet's energy resources. We do not have an absolute right over anything; the trust we have been given must be preserved and handed back to its rightful owner. The principle of accountability acts as a regulatory check against the misuse of God's trust. Absolute sovereignty belongs to God.

The Islamic environment is controlled by the concepts of halal (that which is beneficial) and haram (that which is harmful). Haram includes all that is destructive for an individual, the society, and the environment. An action that brings benefits for an individual but produces concomitant harmful effects for society or the environment becomes a detestable action, and, therefore, one to be outlawed. The conservation and preservation of the natural environment, therefore, features strongly in the Islamic scheme of things.

If other conceptual parameters such as adl (justice), itidal (balance and harmony), istihsan (preference for the better) and istislah (public interest) are added to those already mentioned, then we have the right conceptual ingredients to develop the most sophisticated framework for a contemporary environmental ethic. According to S. Parvez Manzoor,

> Muslim societal ethics, the very basis of society itself, is but a quest for equilibrium, and hence felicity with God, nature and history. It entails submitting oneself to the will of God, accepting the mandate of trusteeship and striving to be a moderate community.
>
> In Sardar, *Touch of Midas*, p. 159

Ziauddin Sardar comments:

> The matrix of this conceptual framework--tawheed, khilafa, amana, halal, haram, adl, itidal, istihsan and istislah--constitutes a paradigm of the Islamic theory of environment. If this framework were fully operationalized in the Muslim umma (community) it would revolutionize the behaviour and thinking of Muslim people. For incorporated in these concepts is a deep respect for nature, an appreciation of interconnectedness of all life, recognition of the unity of creation and the brotherhood of all beings, and that concerns of morality and other living systems must form the basis of any rational inquiry.
>
> *Islamic Futures*, p. 227-8

The concerns we should have for other inhabitants of the biosphere, as well as the environment in general, have been emphasized by Prophet Muhammad, who forbade his followers from harming animals, stressing that the rights of animals should be fulfilled. In the sharia (Islamic law within a framework of ethics encompassing all aspects of living), therefore, animals have legal rights enforceable by the state.

The importance of land reclamation was also emphasized by the Prophet in a number of his traditions. He is reported as having said, for example, that 'whosoever brings dead land to life, for him is a reward in it, and whatever any creature seeking food eats of it shall be reckoned as charity from him'.

Wildlife and natural resources were also protected. Haram, or inviolate, zones bordering watercourses, utilities and towns were established by the Prophet. In such places developmental and construction work is restricted or prohibited. The natural habitat of certain species of flora and fauna is, hence, also protected at the same time.

Hima, or special reserves for safeguarding wildlife and conserving the natural environment, are also an integral aspect of the sharia. It is well known, for example, that the Prophet reserved the surroundings of Medina as a hima for the protection of vegetation and wildlife. He declared as haram (inviolate zones) the private reserves used exclusively by certain individuals. Such sharia-codified principles demonstrate a profound environmental awareness of Islam.

In his conceptual formulation of an Islamic city and the design principles for Islamic environment, Gulzar Haider writes that the city of ecological harmony has

> An evironment that is in a symbiotic, mutually enhancing relationship with nature. Key values are to conserve, not waste; to seek enhancement of the art of life rather than the entropic decay; to seek the ideal that every short-term consumption leads to long-term rejuvenation or creation of new resources; to change the attitude from the survival of the fittest to the survival of the most beneficial. It is an environment that puts great value on physical and mental health. But more than anything else it nurtures the attitude that every situation of apparent conflict between man and nature is an opportunity to design such that the solution is of benefit to both man and nature.
>
> In Sardar, *Touch of Midas*, p. 184

Contemporary Issues: Sexual Morality

Thomas Malthus on Population

Malthus (1766-1834) was a British political economist and is now chiefly remembered for his major work, *Essay on the Principle of Population*, first published in 1798.

In the production of his work, Malthus was reacting against the optimistic ideals of some French philosophers who believed that through the help of suitable education programmes and the due exercise of rationality (power of human reasoning), men might be able to ascend to a Utopian level on earth where all would be bliss. Moreover, they hoped that the French Revolution might be the instrument whereby this could be achieved.

Malthus, however, argued that it would never be possible for such ideals to be realized, for they took no account of an absolutely fundamental feature of society, namely the problem of population growth. Population, he declared, always tends to increase in 'geometric ratio' (e.g. 2, 4, 8, 16, etc.) whereas food production (the means of subsistence) can only be increased in an 'arithmetic ratio' (e.g. 2, 4, 6, 8, 10, etc.). From these premises, Malthus concluded that unless some means of limiting population could be found, mankind must always be subject to famine, poverty, disease and war.

This was a bleak conclusion, indeed, and quite at odds with eighteenth-century optimism. Despite all this, however, Malthus's *Essay* did attempt to offer grounds for hope. The imbalance between population and resources could serve as a spur to activity rather than despair. By their activity, some people at least might gain new qualities and new powers fitting them for a better place in the life hereafter than they could occupy in the present world.

The *Essay* generated much controversy and it appears that Malthus watered down his arguments in later editions. He took the view that rather than be concerned with what might happen in the next world, we should attempt to alleviate suffering in this one. The way to improve human conditions was, he claimed, by the exercise of what he termed 'moral restraint', by which he meant the delay of the average age of marriage, and sexual restraint or abstinence within marriage. Malthus showed, as a result of his observations at home and travels abroad, that most people had some means of limiting population artificially: for example, by celibacy, concubinage, late marriage, infanticide or polyandry. Because man was a reasoning creature, argued Malthus, he was therefore capable of moral restraint in a way that was not open to animals.

Sexual Attitudes of the Permissive Society

We know with the benefit of hindsight that the Malthusian theory, although exercising a powerful influence over the nineteenth-century consciousness, has not been confirmed by events. During the time of Malthus the world population was about 1,000 million. It is now nearly 5,000 million. If the size of the population is to remain constant, it is self-evident that the birth rate and death rate must be equal. The death rate has been continuously decreasing as a result of better diet and improvements in hygiene and medical care. Although the birth rate has been increasing, it has been kept in check in the Western world, particularly since the Second World War, mainly because of improved techniques of contraception.

This technological advance has also, through its abuse, played an instrumental part in bringing about societal decay through profound changes in sexual attitudes and behaviour. Modern societies are no longer in general committed to associating sex with love and faithfulness in marriage. In the mid-twentieth century, sexual 'morality' was severely criticized by an influential minority. It was believed that confining sex to marriage is unnatural and sustained by hypocrisy. The different sexual needs of different people could best be gratified by the use of contraceptives without the risk of conception. Not only had the population theory of Malthus not been confirmed by events, but his sexual morality underpinned by the moral values of Christianity was also jettisoned.

Some researchers began to regard as absurd the idea that

sex should be taught only as part of a loving relationshiop between husband and wife. The sex researcher Martin Cole, for instance, gained publicity for his forthright view that teenagers should be promiscuous, that being promiscuous could be a vitally important part of growing up. He proposed promiscuity camps and the provision of prostitutes on the National Health Service.

The introduction and subsequent misuse of the pill has provided a valuable weapon in the hands of the sexual liberators. Although in theory a method of contraception for married women, it has in practice been used extensively to curb the sharp rise in illegitimate births to girls under sixteen.

In 1972, the Family Planning Association (FPA) declared that the pill 'removes the fear of unplanned pregnancy which might previously have acted as a deterrent to sexual experiment'. It is not surprising, therefore, that since 1974, when contraceptives were authorized under the National Health Service, the Department of Health and Social Security (DHSS) has advised doctors that it is lawful for them to prescribe contraceptives to girls under sixteen. The DHSS has also encouraged doctors not to inform parents without the girl's consent.

As the reigns on sexual morality are loosened, pornography is bound to increase. Pornographic material in magazines, books, films and videos has been defended by its producers on the grounds that it gives pleasure to millions. They have also claimed that by offering an outlet for sexual feelings, they help to reduce sex crimes, although Home Office statistics have shown that reported rape increased by 100 per cent in the decade following the liberation of pornography.

Islam and Sexual Morality

Actions are tied to an individual's perception of his or her destiny. Thus, a person who believes in God and understands his mission as a trustee who will be resurrected after death and held accountable for his deeds in this life, will manifestly behave in a way different from that of a non-believer. Our beliefs are ultimately the basis of our actions.

There is the need for the human to satisfy all aspects of his or her being: the spiritual, intellectual and physical. The sexual need should be considered as one among a totality of needs that collectively produce a healthy body and balanced mind. For the spirit, the believer indulges in remembrance and worship of God. For the intellect there is God's invitation to all human beings that they should ponder on, and understand the natural laws which govern the universe and their own selves. For the physical aspect, believers satisfy their need for food, sex and other material things, but in a wholesome and pure manner without excess. There is no contradiction whatsoever between the satisfaction of sexual needs and the spiritual nature of man. The only impurity comes from the immoral and excessive pursuit of material pleasures. A balance has, therefore, to be struck between the legitimate pursuit of material gain and the pursuit of rewards for the life hereafter.

> Fair in the eyes of men is the love of things they covet: women, children, heaped up hoards of silver. Such are the possessions of this world's life, but in nearness to God is the best of goals to turn to.
>
> Quran 3:14

The pursuit of sexual pleasures regardless of moral considerations has led in our society to widespread adultery, fornication, the overt preoccupation with matters sexual and the development of sex as a business. The other extreme, which particularly existed in the past, is sexual puritanism, according to which sex is viewed as unclean and contrary to goodness, spirituality and faith in God, and should be suppressed and resisted. This view resulted in the institution of monasticism and celibacy, and the belief that marriage and sex were inevitable evils, rather than a blessing from God.

Islam teaches a healthy attitude to sex, encouraging its satisfaction within the bounds of marriage. It prohibits pre- and extra-marital sex as well as homosexuality. There is also an emphasis on etiquette in sexual behaviour between spouses to ensure that there is no selfishness or exploitation.

Muhammad said that whoever marries has already completed half of his or her faith. He also said that Muslim husbands should not go to their wives like animals, but should send a messenger. When asked what is this messenger, he replied, 'the tender kiss or pleasant talk'. Munawar Ahmed Anees has written:

> The moral dilemma in human sexual relations and human reproduction as faced by Western society has no parallel in Islamic history. Compared with the Judaeo-Christian record on human sexuality, Islamic history presents a sex-positive attitude, regulation of matrimonial affairs in a more realistic style, recognition of sexual pleasure within the confines of marital relations, condemnation of celibacy, emphasis on congenial and loving family relations, and due recognition of women as individual human beings in all walks of life. In
> Sardar, *Touch of Midas*, p. 95

By accommodating both sexual pleasure and procreation within the bounds of matrimony, Islam established a just moral order which recognized the innate biological/sexual makeup of human beings. This recognition has been desperately lacking in the modern world and the whole sexual makeup of women has been consistently abused, particularly through the ruthless operation of modern reproductive technologies.

Such technologies have been utilized for the prevention of gametic union mainly outside the institution of marriage. Some of these technologies--mechanical (the condom), chemical (spermicides, oral contraceptive pills) and surgical (vasectomy in men and tubal ligation in women)--have been shown to produce harmful side effects. The occurrence of ovarian cysts and ovarian cancer, for instance, has been linked (as yet without direct correlation) to the use of oral contraceptive pills, and spermicides have been known to produce allergic reactions.

Because of their ready availability, rather than developing in a way so as to alleviate human suffering and help ensure a more stable and balanced married life, contemporary reproductive technologies have served as instruments of the 'sexual revolution' of the 1960s. Their unbridled use and blatant misapplication has directly contributed to the breakdown of moral order, increased licentiousness and promiscuity of the so-called permissive society. This is an example of societal immorality developing and evolving within the matrix of an unhealthy civilizational worldview. The scientific knowledge and technology is there, but where is the institutionalized system of ethics and morality within the framework of which it should permanently operate?

The Information Revolution

A professor of science education said that the greatest polluter of the twenty-first century is going to be information. We as citizens of the modern world are constantly bombarded by all sorts of information, perpetually churned out and mercilessly inundating our institutions.

The information revolution is, of course, a mixed blessing. On the one hand, the profusion of information technologies has meant that a vast amount of work can now be done electronically which at one time was unimaginable. On the other hand, and increasingly more worrying, is the fact that the information age, far from increasing our control over our own lives produces the opposite effect.

The ethics of Islam can provide a solution to the deterministic aspect of the information age. If the Muslim world mistakenly were to adopt such a totalitarian technology as that spreading through the Western world, they would risk inducing a new, more subversive and devastating type of dependency, a neo-colonialism of the information age. If, on the other hand, they completely ignore the developments in informtion technology, they risk leaving their destiny in Western hands. As information technologies are becoming the basic tools of manipulation and control, access to them will become the decisive factor between control and power or manipulation and subservience. In this powerful dilemma, the way forward, surely, is to modify the technology at the point of use to suit the needs and requirements--the goals--of Muslim society. All such modification must, of course, be done in accordance with the ethical dictates of Islam. The deterministic consequences of a world based on linear logic must be altogether avoided. Meaning and purpose are lost in such a world, human beings becoming mere automata to be controlled and manipulated by the social engineers and power brokers of the day.

> The Brave New World of AI [artificial intelligence] is ruled by the demigod of automation. Man need not do what may be accomplished by the machines, seems to be inscribed in its holy creed. In this narcissistic universe of self-gratification, there is no room for self-sacrifice. Little wonder that the instrumentalist approach to thinking and the computational approach to intelligence have today triumphed over the human-centred ideals of self-transcendence. Today, the technical mind is making its impudent assault on the last bastion of human autonomy--the human mind. If man earlier could take comfort in his being free in mind and thought, he has reasons to be apprehensive today. What AI aims at is not the simulation but the replacement of human thought.
>
> Manzoor, p. 39

Islamic Science: The New Paradigm

The cardinal concept of Islam--tawheed--is violated by the deterministic aspects of a society controlled by information technology. God is not required in a society where everything can be predicted by computers. The ever-increasing super-processing power of central processing units undermines, by extension, the notion of khilafa, or the role of human beings as the representatives of God. Furthermore, the promotion of computer-oriented rather than labour-intensive production denies adl (social justice) to the people, those with access to large data bases being able to manipulate the citizens socially and economically. This, of course, is against the public interest, which also is not served by the invasions of privacy that are the consequence of institutional desire to gather as much information about the personal affairs of citizens as possible. In 'Choruses from "The Rock"', T.S. Eliot wrote, 'Where is the wisdom we have lost in knowledge? Where is the knowledge we have lost in informtion?' He was referring to the process of acquiring knowledge being reduced to the clinical manipulation of information. From the Islamic point of view, in such cases ilm ceases to be a virtue, and therefore ibadah (worship) becomes little more than a mechanical exercise. This is the danger of linear, deterministic thinking, which computerized systems promote. Hence, the fundamental Islamic notion of hikma (wisdom) is also violated as the holistic perspective to human problems (the complex interconnection of the process of living) is not brought in to play.

Islamic values, therefore, are of great contemporary importance in developing information policies and strategies for the twenty-first century. Issues concerning industrial automation, development of large data banks containing personal information about citizens, and the potential invasion of privacy through the connection of various data banks, are all pressing problems for the modern day world, for which Islamic values may provide the basis for solutions. Islam abhors the use of information technologies to control, manage and manipulate human thought and action for selfish personal or institutional reasons. Yet this is what is happening on a large scale in the contemporary Western world. In the conceptual matrix of Islam we find the intellectual framework for a humane and lasting solution.

Conclusion

The practice of Islamic science creates an atmosphere that encourages the remembrance of Allah, motivates behaviour according to the dictates of the sharia and promotes the conceptual values inherent in the Quran. It is a living, dynamic entity able to provide contemporary solutions to contemporary problems within the most humane and ethical framework in perfect harmony with man and nature. It is a science truly international in character, for the re-establishment of which a major intellectual revival is not only desirable but necessary.

> Islamic science came into being from a wedding between the spirit that issued from the quranic revelation and the existing sciences of various civilizations which Islam inherited and which it transmuted through its spiritual power into a new substance, at once different from and continuous with what had existed before it. The international and cosmopolitan nature of Islamic civilization, derived from the universal character of the Islamic revelation and reflected in the geographical spread of the Islamic world (dar al-Islam), enabled it to create the first science of a truly international nature in human history.
> Nasr, *Islamic Science*, p. 9

3

Islamic Science in History

> Say, shall those who have knowledge and those who have it not be deemed equal?
> Quran 39:12

It can be confidently said that all Muslim intellectual activity in the widest sense had its genesis in the Quran. Muslim scientists had a great thirst for knowledge which was underpinned by their religious fervour. Whether astronomers, mathematicians, chemists or physicians, they sought to work for the glory of God and the service of Islam as they assimilated and then synthesized the sciences derived from Greece, Persia and India. The Quran was their supreme inspiration.

> We will show them the signs in all regions of the earth and in themselves until they come to see that this is the truth.
> Quran 41:53

The endeavour to develop Islamic science was helped a great deal by contact with other nations and cultures, such as Syria and its Byzantine culture, with Persia of the Sasanids, with India, and subsequently with North Africa and Spain. Many of these peoples embraced Islam. Others, 'the people of the book' (Christians, Jews and Sabaeans) were given the protection of Islamic law as dhimmis (non-Muslims living in a Muslim state) and encouraged to take an active part in scientific and cultural life. They all contributed to the development of sciences in Islam and wrote their works in Arabic which became acknowledged as the language of scientific communication.

Islamic science in history is a vast and complex enterprise that took root in the early Abbasid period at Baghdad shortly after 750 CE and thrived for about six hundred years thereafter. During this time it spread over a vast geographical area that extended from Andalusia to Central Asia.

It is tempting to assume, as some writers have done, that Islamic science is the continuation of a Greek tradition that had been preserved by the Hellanized peoples who came under Arab rule. Many historians, however, have since noted that it was much more than that. Baghdad was heir not only to Alexandria but also to Persia and India. It is a remarkable fact that scientific texts available for translation into Arabic included works in Syriac, Sanskrit and Pahari, as well as in Greek. Such assimilation and subsequent synthesis led to the accumulation of a body of scientific learning that surpassed anything previously known.

As a result of this learning and its subsequent dissemination, numerous Arabic words have passed into some of the Western languages, especially terms used in chemistry, navigation and astronomy. Algebra, alcohol, alchemy, alembic, alkali, azimuth and zenith are examples. This in itself is eloquent testimony to the fact that Arabic, which had been the language of poetry and of the Quran, became the international language of science. In fact, by the eleventh century, the great Persian scientist al-Biruni was describing the Arabic language as the language most suited for scientific expression. Such was the position of Islamic science.

The individuals who contributed to the growth of Islamic science represented many ethnic groups and, at least in the beginning, professed different faiths. For example, many Syrian Christians and Persians were among the first translators, physicians and astronomers attached to the Abbasid court. Some, such as Thabit bin Qurrah, were Sabaeans from pagan Harran; others, such as Masha Allah, were Jews. In the later history of Islamic science some of the most important figures came from such non-Arab lands as Khwarizm, Farghana, Sijistan, Khurasan and Fars. The unifying thread was the Arabic language, a fact of great historical significance. Islam provided the right intellectual atmosphere in which non-Muslims could work and contribute positively. The illustrious British philosopher Bertrand Russell wrote:

> Throughout the Middle Ages, the Muslims were more civilised and more humane than the Christians. Christians persecuted Jews, especially at times of religious excitement; the Crusades were associated with appalling pograms. In Muslim countries, on the contrary, Jews at most times were not in any way ill treated. Especially in Moorish Spain, they contributed to learning; Maimonides (1135-1204) who was born at Cordova, is regarded by some as the source of much of Spinoza's philosophy. When the Christians reconquered Spain, it was largely the Jews who transmitted to them the learning of the Moors. Learned Jews, who knew Hebrew, Greek and Arabic and were acquainted with the philosophy of Aristotle, imparted their knowledge to less learned schoolmen. They transmitted also less desirable things, such as alchemy and astrology.
>
> Russell, p. 324

An efficient form of epistemological classification also helped the development of Islamic science by putting knowledge in an appropriate social and ethical context. The great Muslim thinker, al-Ghazali (1058-1111) was the most significant writer in this regard. He gave a detailed classification of the sciences from the standpoint of moral and legal obligation. Certain sciences, such as law and ethics, are compulsory for each individual (fard ayn), while such others as medicine and agriculture are compulsory only for the community (fard kifaya). The knowledge of revelation communicated by the prophets transcends these epistemological divisions and cannot be acquired by reasoning (like arithmetic), by experiment (like medicine) or by ear (like language).

The non-religious sciences may be recommended, may merely be permissible, or may be culpable or blameworthy. Recommended or praiseworthy sciences are those without which life in a community becomes impossible, such as medicine and agriculture. It is therefore a communal obligation to develop them. Permissible sciences have an optional character but their development is a laudable enterprise since it increases the competence of the community. The pursuit of engineering or medicine beyond the point required by practical utility would be a commendable act if the community possesses the means to do so. Culpable sciences, such as magic, astrology and (in the modern world) certain forms of genetic engineering as well as the scientific study of torture, are to be thoroughly condemned and never allowed to take root in society. Scientific knowledge is thus placed in an appropriate social and ethical context, subordinated to the knowledge of revelation and subservient to the aims and objectives of the Muslim community.

The Physical Sciences

Mathematics

Muslim scientists developed mathematics and used it to solve such problems of daily life as the assessment of taxes, reckoning of legal compensation and division of inheritance according to Islamic law. Taking basic principles and definitions from the Greek, they developed the science of mathematics theoretically and practically. The applications of their mathematics included the study of problems of, for instance, surveying, the construction of improved mills, the

Islamic Science in History

study of mechanical tools, and the introduction of wheels with scoops for the continuous drawing of water from a watercourse.

In developing mathematical theory, they used exponential terms, and the extraction of square and cube roots. They were aware of algebraic/arithmetical laws such as '$an + bn = (a + b)n$'. To make calculations easier, they constructed abaci. Their blend of enthusiasm and ingenuity led them to remarkable discoveries. The Sabaean, Thabit bin Qurrah (d. 901), for example, discovered the characteristics of amical numbers: those in which the sum of the proportional parts of one is equal to the other and vice versa. For example, 220 and 284 are amical numbers because $220 = 1 + 2 + 4 + 71 + 142$ (the parts of 284), and $284 = 1 + 2 + 4 + 5 + 10 + 11 + 20 + 22 + 44 + 55 + 110$ (the parts of 220).

A superior system of reckoning came from India, which was able to express any number, however large, by means of nine figures and a symbol: the zero or sifr. Muslim scientists in Arabia used this system and elaborated it, whence it was transmitted to the European world via a handbook written by Muhammad bin Musa al-Khwarizmi, who flourished around 825 CE. The Latin version of his treatise is extant today.

Al-Khwarizmi was also the first author in the field of Muslim algebra. In fact, the very word 'algebra' comes from the Arabic term al-jabr, which was one of the mathematical operations al-Khwarizmi used for solving quadratic equations. The translation in the latter half of the twelfth century of the first part of al-Khwarizmi's *Kitab al-Jabr wa al-Muqabalah*, first acquainted the European world with Muslim algebra. By this time, however, the Muslims were well advanced in their solutions of geometrical problems suggested by Greek mathematicians. Such mathematicians as al-Karaji, who flourished around 1000 CE, Umar al-Khayyami (d.1130), ibn al-Haytham (d.1040) and al-Samawal (d.1175), for example, led the Muslim world to mathematical successes during the Dark Ages of Europe.

Astronomy

Astronomy was closely connected to prescribed religious practices. Instruments and computational techniques were developed to determine the hours of the day and night in order to establish the times of the five obligatory prayers. Muslim astronomers carefully studied the work of the Greek astronomer Ptolemy, as expounded in the *Almagest*, and then made it more accurate and efficient. They developed ingenious computational techniques for determining planetary orbits, with applications to the composition of astronomical tables and the theory of instruments. Their planetary models and observational techniques improved Ptolemaic parameters such as mean planetary motions, precession of the equinox, and inclination of the ecliptic.

It was the Caliph Abdallah al-Mamun (813-33) who gave a strong impetus to observational astronomy by ordering the preparation of new astronomical tables. For this purpose, observational and computational instruments such as astrolabes and quadrants were used at the various centres of astronomical research in the Islamic world. The precision and sophistication of these instruments was such that, for example, the description of an instrument constructed in the thirteenth century at Maragha by the Damascene astronomer al-Urdi has been compared with a similar one made and used in the sixteenth century by Tycho Brahe.

Classical Muslim astronomy developed along several distinct lines, influenced by the example of the different models presented in the translated works. The works of Aristotle and his Greek commentators had acquainted the Arabs with the Eudoxian model, a homocentric system in which each planet was associated with a number of spheres that revolved around the centre of the world, believed to be the Earth. Aristotelian and Ptolemaic astronomy had developed this idea of geocentrism (Earth at the centre of the universe) to which the Muslim astronomers became heir. At that time heliocentrism (the system where the sun occupies the centre of the solar system) could not be irrefutably demonstrated, nor, in the absence of the telescope, could it be of any use in practical astronomy.

Aristotle had adopted the geocentric model and developed a rigid cosmological theory from it. He believed that celestial bodies were perfect and incorruptible. He maintained that they rotated in perfect circles with constant speed, were transparent, and that there was only one centre for all heavenly motions. Despite his tremendous prestige, many Muslim astronomers did not adhere to Aristotle's model. Not, that is, until twelfth-century Muslim Spain when philosophers such as ibn Rushd, Maimonides and ibn Tufayl revolted against the entire Ptolemaic system and demanded a return to a model in closer agreement with Aristotelian cosmology.

Muslim astronomers knew of the same planets and their movements as were known to the Greeks. Moreover, their method of representing them was also similar, subscribing to the principle of uniform circular motion for celestial bodies but admitting eccentric and epicyclic motions. The sun, for example, moved in a circle eccentric to the centre of the world. The other planets moved in little circles, called 'epicycles' by Ptolemy, which were themselves carried around large circles called 'deferents'. Some Muslim astronomers, however, rejected this theory because they felt that some 'physical' proofs had crept into Ptolemy's otherwise mathematical exposition. This was in connection with a long-running debate about the relation, or lack of relation, between physics and mathematical astronomy. The majority of Muslim astronomers, in fact, believed that astronomy was based upon physical as well as mathematical premises, and so they approved of Ptolemy's planetary model formulated in the language of arithmetic and geometry with some physical explanations.

Subsequent Muslim astronomers led by Nasir ad-Din al-

Tusi at Maragha initiated important reforms of planetary astronomy which were continued later by Ibn ash-Shatir at Damascus. The successful planetary models they produced were compatible with contemporary astronomical principles: that the apparent motion of a planet be represented as the resultant of a combination of motions, each being uniform with respect to its own centre. Historians of science have noted a strong similarity between the planetary models produced at Maragha and Damascus and those of Copernicus, the Polish cleric who revolutionized Western planetary astronomy in the sixteenth century. Any causal connection, however, remains a matter of conjecture.

Optics

Al-Hasan bin al-Haytham (d. 1039), well known to the Western world as Alhazen, may be considered to be the father of optics. Building on the geometrical foundations laid by the ancient Greeks, he developed a theory of optics that was to have profound influence upon the work of many Western scientists. His book, *Kitab al-Manazir* (On Optics), for example, exercised an important influence in the Middle Ages, prompting the studies of Roger Bacon.

Ptolemy's treatise on optics, the most mature work on the subject produced in antiquity, was therefore a treatise of Greek optics that was primarily a theory of vision. Al-Haytham produced a theory of vision quite distinct from that of the Greeks, and presented it with physical as well as mathematical explanations. Using Euclidean and Ptolemaic geometry, he derived the idea that vision occurs when light emanates from an object and enters the eye. This prompted him to study the visual impressions created in the eye and brain by both small and large objects. His study led him to formulate an original theory of the psychology of visual perception which was expounded in his *Book of Optics*.

Islamic Science in History

Applying his knowledge of reflection and refraction, al-Haytham investigated the phenomenon of atmospheric refraction, calculating the height of the atmosphere (10 miles). In his study of lenses, he experimented with different mirrors--flat, spherical, concave and convex, parabolic and cylindrical--from which he originated important discussions of the rectilinear propagation, reflection, and refraction of light and colour.

The first known instance of the camera obscura may be attributed to al-Haytham, after he made observations of the semi-lunar form of the sun's image cast during eclipses on a wall set opposite a screen with a tiny hole in it. He also solved wholly geometrically a physical problem known as al-Haytham's problem, namely, to find the reflection point, given the object and image by reflection in a spherical mirror. It was not until the mid-seventeenth century (about 600 years later) that Huygens and Sluse discovered the algebraic solution to this problem.

It soon became clear to Latin medieval writers that Ibn al-Haytham's *Book of Optics* was superior in its analysis and presentation to the treatises of Euclid, Ptolemy, al-Kindi and Ibn Sina, all of whose work had also been translated into Latin. In thirteenth-century England, Roger Bacon often referred to Ibn al-Haytham as 'the author on optics', when other writers, particularly Witelo, actually made use of a substantial amount of al-Haytham's text on optics to produce their own works on the subject. Even then, they did not fully appreciate the mathematical aspect of al-Haytham's work which was not comprehensively understood in the West until the seventeenth century.

In the Islamic world, for reasons that are not quite clear, Ibn al-Haytham's text did not receive the attention it deserved until the end of the thirteenth century, when one of al-Haytham's successors, the Persian Kamal ad-Din al-Farisi (d. 1320), wrote a critical commentary of the Arabic text.

Al-Farisi repeated and improved the accuracy of the

experiments of al-Haytham on the camera obscura and also put forward a viable explanation of the rainbow phenomenon, which had resisted the efforts of all his predecessors since antiquity. Inspired by Ibn Sina's analogy between a raindrop and a glass sphere, al-Farisi observed the path of light rays in the interior of a glass sphere, hoping to determine the refraction of solar light through raindrops. He was thus able to give a valid explanation of the formation of the primary and secondary rainbows. The most impressive thing about his, as well as Ibn al-Haytham's work, is their experimental approach to scientific problem-solving with the application of mathematical theory in their explanation of the problem.

Chemistry

When Muslim scientists first translated the chemical books of the Greeks into Arabic they discovered that Greek chemical thought was mixed with mythical aims. The philosopher's stone which converts base metals into gold and the elixir of life that permits eternal health and youthfulness were the primary objectives of Greek chemical research. The Muslims, in fact, were the first to test the value of different chemical theories by experimentation. They discovered many new products on the way, the names of which in European languages still sound similar to the original Arabic. For example:

    chemistry - al-kimiya
    alcohol   - al-kuhul
    alkaline - al-qalawi
    arsenic   - al-zirnich

Probably the greatest Muslim chemist was Jabir Ibn Hayyan (738-813) from Kufah in Iraq. He experimented on material of animal, vegetable and metallic origin, devising

Islamic Science in History

instruments for cutting, calcining and crystallizing in the process. He described and perfected the basic processes of sublimation, evaporation, liquefaction, crystallization, calcination, amalgamation, ceration (waxing), oxidation and purification. Ibn Hayyan claimed that water can be pure only by distillation, differentiating between direct distillation using a wet bath and indirect distillation using a sand bath.

Many new chemical substances were identified by this great chemist, including alkalis, acids, salts, paints and oils. He prepared sulphuric acid, caustic soda and nitro-hydrochloric acid. (The latter is known as aqua regia or royal water, and is a mixture of nitric and hydrochloric acids used to dissolve metals such as platinum and gold.) He also produced ethanol acid (acetic acid, which he called vinegar acid), and numerous salts such as sulphates, nitrates and both potassium and sodium carbonates. More practically, he prepared paints of different colours for use on clothes and animal skins, as well as an ink to be used for expensive manuscripts.

Although Ibn Hayyan wrote more than 500 studies in chemistry, few of these have reached us. As an example of the greatness of his chemical intellect, a perceptive passage is quoted below from one of his extant works:

> Mercury and sulphur unite to form one single product, but it is wrong to assume this product to be entirely new and that mercury and sulphur changed completely. The truth is that both kept their natural characteristics and that all that happened is that parts of the two materials interreacted and mixed, in a way that it became impossible to differentiate them with accuracy. If we were to separate...the tiniest parts of the two categories by some special instrument, it would have been clear that each element kept its own theoretical characteristics. The result is that the chemical combination between the elements occurs by permanent linking without change in their characteristics.

The special instrument imaginatively referred to by Ibn

Hayyan is now used by chemical research laboratories throughout the world. It is the mass spectrometer, which ionizes gaseous atoms and molecules and identifies them on the basis of their different mass.

The Muslim chemist from Persia, Al-Razi (824-932) built upon Ibn Hayyan's work and perfected the process of experimentation by describing first the materials he used, then the apparatus, methods and conditions of the experiment. He also prepared sulphuric and other acids as well as alcohol by the fermentation of sweet products, studied mercury and its compounds and described the design and use of a number of instruments for use in chemistry. Al-Razi was the first to divide chemical substances into the categories of mineral, vegetable and animal, as well as declaring that the functioning of the human body is based upon complex chemical reactions.

Consider now a quotation from one of the works of Al-Majriti, the Andalusian chemist from Majrit (today's Madrid):

> I took clean, shining mercury and I put it in an egg-shaped utensil made of glass, and I introduced it into another utensil similar to kitchen utensils. I let it warm under such a low fire as I could put my hand on the outer surface of the instrument. The heating continued for forty days, and when I opened the instrument I found the mercury (which weighed a quarter of a pound) had been transformed into a red powder without any change in the overall weight.

The red powder was, of course, mercury oxide. Although Al-Majriti first proved the principle of chemical conservation of mass, credit for this was given 900 years later to the French chemist Lavoisier, who performed a similar experiment. Muslim chemistry was thus centuries ahead of its time.

Islamic Science in History

The Life Sciences

Zoology and Botany

The nomadic or semi-nomadic existence of the tribes of the Arabian peninsula, faced with harsh environmental conditions, necessitated a reliance on domesticated animals for survival. This, together with the quranic prescription to study the signs of God through the study of nature, was the basis for the strong interest of Muslims in the care and feeding of animals for food, by-products and transportation.

As far as serious intellectual studies are concerned, it was al-Jahiz (d. 869) who wrote *al-Hayawan*, the first comprehensive zoological study of animals in the Arabic language. It is a compendium of zoology and veterinary medicine, describing animal types, their characteristics and behaviour, diseases and treatment in what is modern-day Iraq and neighbouring countries.

Some form of zoological taxonomy was attempted by the Egyptian philosopher Kamal ad-Din ad-Damiri (d. 1405) in his monumental work *Hayat al-Hayawan* (The Life of Animals). He listed the characteristics of animals by arranging them in alphabetical order, and discussed the medicinal value of their organs.

One of the most comprehensive works of zoology and veterinary medicine was produced by Abu Batr al-Baytar of Cairo (d. 1340). In his manual *Kamil as-Sina'a tayn*, al-Baytar covers a variety of topics: animal husbandry, breeding, variations in wild and domestic animals, as well as a section on ornithology, in particular a study of those birds domesticated in Egypt and Syria. A substantial part of his work is also devoted to a discussion of animal diseases, the methods and drugs used in their treatment, as well as the use of animal organs in therapeutics.

Just as the knowledge of zoology was used to develop the science of animal husbandry, botanical knowledge was applied in the development of agricultural science. As Islam expanded, agricultural and horticultural activities flourished, and several detailed works were written in Arabic throughout the Islamic domain, from the eastern regions to Andalusia.

Two books written in Muslim Andalusia deserve special mention. One was completed in the latter half of the eleventh century by Ibn al-Bassal of Toledo, re-edited with a Spanish translation this century under the title *Libro de agricultura*. The other highly regarded text on the subject was the twelfth-century *Kitab al-Filahah*, written by Ibn al-Awwam of Seville. Both are detailed texts covering topics such as plant taxonomy, soil, farming techniques, methods of cultivation, irrigation, tillage, gardening and landscaping, agronomy, medicinal plants, and plant reproduction.

In addition to Muslim Spain, botanical and agricultural science flourished in Syria, Iraq and Egypt during the same period and continued until the end of the fourteenth century.

Medicine, Surgery and Medical Education

Medicine had a long history before it was developed to a remarkably high level by the Muslims. Hunayn bin Ishaq al-Ibadi (809-873) did Islamic medicine a great service by translating into Arabic with the help of his team of able translators the most important medical works of the Greeks, including the writings of Hippocrates, Dioscorides, Galen and the doctors of the school of Alexandria. Muslim medics subsequently built a towering edifice upon this intellectual foundation.

Apart from translations, the indefatigable Hunayn also wrote one hundred or so works himself, mainly concerned with medicine. Those of his books which had the most influence in the Western world were *Medical Questions*, a general introduction to medicine in the form of questions and answers, and two works on ophthalmology, *Ten Dissertations on the Eye* and *Questions on the Eye*. In the latter works, Hunayn explained the anatomy of the eye, widening it to a physiological description of the brain and optic nerve. Moreover, he examined nosology, aetiology and symptomatology, not to mention diseases of the eye and the properties of useful medicaments. He punctuated the text with excellent diagrams of the anatomy of the eye, much superior to those produced in the Middle Ages by the medics of the West.

Without doubt, the greatest Muslim physician was Abu Bakr Muhammad bin Zakariya ar-Razi (865-925). As well as being a great clinical doctor, he was also an eminent pathologist, medical eductor and philosopher of his time. His literary output in all these areas of learning was encyclopaedic. According to al-Biruni, ar-Razi wrote fifty-six medical treatises, thirty-three on natural sciences, twenty-two on chemistry, seventeen on philosophy, fourteen on theology, ten on mathematics, eight on logic, six on metaphysics and a further ten on miscellaneous subjects: an astounding 176 books.

So great was ar-Razi's contribution to medicine and so powerful his academic impact that many of his original ideas and concepts remain valid today. He produced influential writings on the doctor-patient relationship, the diagnosis of diseases, psychiatrics, chemotherapy and therapeutics. He gained international recognition as a result of his discourse on smallpox and measles. Dealing with the causes, diagnosis and treatment of smallpox, ar-Razi was able to produce a list of characteristics that differentiated the disease from measles. In identifying the specific symptoms of smallpox and measles, he used concepts that are accepted in contemporary pathology.

Perhaps not surprisingly then, the work which deals with smallpox and measles is the most famous of ar-Razi's medical works. Known in the medieval Latin translations as *De variolis et morbilis*, the book is truly original, based on ar-Razi's personal observations and clinical deductions. In fact, it is the first treatise in existence on infectious diseases.

Ar-Razi's other most celebrated work is his *Kitab al-Hawi Fil-tibb* which in Latin is known as *Contineus*, that is to say a work containing the whole of medicine. In this comprehensive medical encyclopaedia he discusses the methods, applications, and scope of internal, clinical and psychiatric medicine, and attempts to rationalize the relationship between psyche and soma. Other subjects meriting his analysis include diets and drugs and their effects on the body, a regimen for preserving health, skin diseases, mouth hygiene, the effect of environment on health, epidemiology, toxicology, as well as general medical theories and definitions--a truly comprehensive treatise.

Some of ar-Razi's illustrious successors included Ibn al-Jazzar (d. 984) who wrote on therapeutics, dietetics and internal medicine; Ali bin Abbas al-Majusi (d. 994) who studied, among other things, the impact of environment upon health, the nutritional value of diets, and the action of drugs on human beings; and the physician-philosopher al-Mukhtar bin Abdun bin Butlan (d. 1068) who wrote extensively on health, postulating certain medical principles that need to be kept in balance in order for the body to maintain a state of equilibrium. The works of these Muslim scholars was translated into Latin and received much attention in European medical circles of the Middle Ages, surpassed only by *Al-Qanun Fil-tibb* (the Canon of Medicine) by the celebrated physician-philosopher Ibn Sina, who was known in the West by his Latinized name, Avicenna (980-1037).

With Ibn Sina, Islamic medicine reached the acme of its achievement. Ibn Sina successfully attempted a rationalization of the immense accumulation of medical science inherited from antiquity and enriched by his predecessors. The *Canon of Medicine* consisted of five books, the most complete dissertations of their time. He begins by expounding the generalities of medical science, such as the anatomy of various organs and tissues, diseases and their causes, hygiene, and general rules for treatment. He describes all known medicaments of animal, vegetable and mineral origin. Disorders of the body, particularly those affecting each limb, are described and classified from head to foot in descending order. There is then some discussion of fevers, tumours and pustules, poisons and fractured limbs, followed by an exposition of the nature and action of compounded medicaments such as powders, dry drugs, potions and syrups. Ibn Sina's *Canon*, which was enthusiastically studied and lavishly illustrated by subsequent Muslim physicians who also made summaries of it, won him great prestige in medical circles during the Middle Ages and after.

As a further development in the practical application of medical knowledge, Muslim surgeons were among the first to use narcotic and sedative drugs in operations. The physician-philosopher Ibn Rushd (1125-98), known in the West as Averroes, once said that 'whoever becomes fully familiar with human anatomy and physiology, his faith in Allah will increase.' This perceptive statement gives a good indication of the context in which Muslims practised their science.

Building on the translated works of Galen on anatomy and surgery, al-Majusi wrote his influential book *Liber regius*, the first Islamic work to deal with surgery in detail. Al-Majusi was the first to use the tourniquet to prevent arterial bleeding.

The greatest achievements in Muslim surgery, however, may be attributed to az-Zahrawi of Moorish Spain (940-1013). In his encyclopaedic work, *at-Tasrif*, he dealt with obstetrics, paediatrics and midwifery, as well as described many surgical techniques, including cauterization, the setting of bones in simple and compound fractures, treatment of wounds and techniques to widen urinary passages for the surgical exploration of body cavities. He also designed many surgical instruments which became the prototypes upon which many subsequent surgical instruments of Western medicine were based.

In terms of institutionalizing medicine and medical education, hospitals flourished throughout the Muslim world under the patronage of the Muslim caliphs. Once again, as in other areas of science and its applications, Muslim hospitals became the prototype for the development of the modern hospital. From the ninth century onward, institutions were established devoted to the promotion of health, cure of diseases, and the teaching and expanding of medical knowledge, serving men and women in separate wards. A case in point is the Adudi hospital in Baghdad: built in the late tenth century, it had twenty-four doctors on its staff and was equipped with lecture halls and library. It became a centre of excellence for aspiring medical students throughout the Muslim world, many of whom came from far away in order to study there. The Adudi hospital set the pace for the subsequent momentum in hospital construction and medical education. Muslim medical establishments attracted students from the West as well as the Muslim world, enjoying generous endowments from the state treasuries. It is reported that the buildings were spacious, equipped with comfortable lecture halls, large libraries and good laboratories. There were even well-stocked pharmacy shops where freshly prepared medication could be dispensed. This was the undoubtedly the golden age of Muslim civilization.

Pharmacy and Pharmacology

Alongside medicine, yet having its independent identity, pharmacy developed under Islam, particularly in the early ninth century in Baghdad, the Abbasid capital. Here, drugs and spices from Asia and Africa were readily available and the need for medication was given added emphasis due to the proximity of military installations.

In addition to the preparation and dispensing of pharmaceutical products in Muslim hospitals, pharmaceutical preparations were also commercially distributed in the marketplace and dispensed by physicians and pharmacists in forms ranging from pills, ointments and electuaries, to tinctures, suppositories and inhalations. The science of pharmacy was, indeed, most advanced for its time.

Among the pharmaceutical texts that achieved popularity were the treatise on pharmacy by ar-Razi and Books Two and Five of Ibn Sina's *Canon*. The most important text on pharmacy and pharmacology (which developed in close connection with the former), however, was the *Kitab as-Saydalah Fil-tibb* (the Science of Drugs) by Abu ar-Rayhan al-Biruni (d. 1051). In this epic work, al-Biruni gave a professional exposition of pharmacy and the responsibilities of the pharmacist, and also developed a detailed definition of pharmacology.

Muslim pharmacists and pharmacologists enriched the *materia medica* inherited from the Greeks by adding many valuable remedies such as camphor, tamarind, nutmeg, senna, ergot, rhubarb, and a host of other (now obsolete) drugs. They gave detailed descriptions of the drugs they discovered with reference to their geographical origins, physical properties and methods of application. They also experimented with drugs in order to learn more about their effects on human beings, basing their treatment of certain ailments on their empirical findings. Pharmacological manuals of the time even included quantitative information in relation to the potency of drugs and recommended dosages according to age, sex and degree of illness.

Islamic pharmacy and pharmacology was based on the Islamic teaching that Allah has provided us with a great variety of natural remedies to cure our ills. It is our duty to identify and utilize them skilfully and ethically. Interest in natural products and ecology was, therefore, a natural corollary to this Muslim belief. Natural medications were seen as tokens of Allah's magnanimous attitude towards human beings, which is why Muslim pharmacists, physicians and pharmacologists sought remedies in nature, rather than in laboratory-prepared synthetic drugs.

The Spirit of Islamic Science

The special features of Islamic science in history included a high level of polymathy, and intense external and internal criticism, together with a deep connection of the enterprise with society and Islamic institutions. Determination of the direction of the *qiblah* for prayer, the study of the new moon in connection with Ramadan and the subsequent development of lunar astronomy, and the development of geography for the purposes of making pilgrimages more convenient are all examples of the unique spirit of Islamic science in history.

Mathematics in Islam developed directly from the realization that the multiplicity of the physical and biological world is a manifestation of divine unity, powerfully shown in Islamic art and architecture--from patterns on carpets to the ornaments of mosques--where geometry and arithmetic come together to produce kaleidoscopic designs that reflect the unity of the Creator and the multiplicity of the created.

> Lo! In the creation of the heavens and the earth, and the difference of night and day...are signs (of Allah's sovereignty) for people who have sense.   Quran 2:164

Verses such as this provided a strong intellectual impetus to Muslim scientists to study astronomy and cosmology, the practical impetus coming from the importance to the religious community of the five daily prayers. The timings of the daily prayers had to be determined throughout the year, as did the direction for the prayers facing Makkah. Subsequently, as Islam spread far and wide, the timings of prayer and the direction of Makkah (the qiblah) had to be determined for all global localities where believers were resident. As necessity is the mother of invention, religious necessity was the impetus for Muslim astronomers such as al-Biruni and al-Haytham to devise instru-mental means for finding the direction of Makkah and mathematical methods for calculating its direction. Intellectual needs associated with the nature of the quranic revelation therefore combined with practical religious needs to make astronomy a primary concern of Muslim scientists.

The most important Islamic astronomical instrument developed as a result of the fulfilment of such intellectual and religious needs is, of course, the astrolabe. Some of these instruments are outstanding examples of Islamic art, having caught the attention of Europeans in the nineteenth century. The instrument itself is multifunctional, used to determine the altitude of the stars, the sun, the moon and other planets in much the same way as the sextant or quadrant, as well as used to tell the time and to measure the height of mountains and the depth of wells. Muslim scientists also made numerous zodiacal armilleries and celestial globes to enable them to predict planetary longitudes for religious purposes. Many of their fine instruments are housed today in European and American museums.

Muslim studies in geography and geodesy drew from many sources, such as Babylonia, Persia (in particular), Greece and India. Makkah being the centre point for the final revelation from Allah and the focal point for the annual Hajj (the most powerful manifestation of the internationalism of Islam), the Islamic geographers devised a geographical worldview in which the central region of the world was conceived in a new way so as to encompass Makkah. Quranic revelation encouraged the study of descriptive and quantitative geography, considered as 'scientific' geography today. Seyyed Hossein Nasr has put it very well:

> As far as geodesy is concerned the Islamic geographers made major contributions to it as well, and in fact al-Biruni may be

> considered as the founder of the science of geodesy. Muslims were interested in mathematical studies of the features of the surface of the earth as well as in determining the latitude and longitude of cities, the height of mountains, the diameter of the earth, etc., from the moment they inherited the Greek, Indian and Persian works on the subject. But they were also interested in some of these problems for the practical purposes of orientating themselves in the direction of Makkah and finding the various times of day in order to perform the prescribed prayers and to fast properly. Treatises on how to determine the length of day abound in Islamic languages, as do works on ways of determining the direction of Makkah from various latitudes and longitudes.
>
> Nasr, *Islamic Science*, p. 48

To take further examples of this unique spirit from the biological sciences, Islamic medicine was more concerned with the prevention of illness than with its cure. Consequently, the question of hygiene and preventive medicine played a major part in both theory and practice. Western travellers to the Islamic world were often struck by the general cleanliness of the people.

The teachings of Islam emphasized the importance of personal hygiene and cleanliness. The regular washing for prayer, use of the toothbrush which goes back to the Prophet, and strict injunctions concerning private and public hygiene are rooted in the teachings of Islam. What are regarded as excellent dietary habits today, such as eating less than one's full appetite and eating slowly, were, in fact, part and parcel of Islamic religious teachings, and therefore formed the theoretical basis of Islamic medicine. Total abstention from alcoholic drinks and pork, and also fasting, were other dietary habits, discussed within the framework of hygiene and public health but rooted directly in the teachings of the sharia (Islamic law encompassing ethics and methodology).

The Muslims considered the kind of food and the manner in

which it is consumed to be so directly connected to health that the effect of diet was considered by them as being perhaps more powerful than that even of drugs on both health and illness.

Nasr, *Islamic Science*, p. 166

As a result, many works were entitled *hifz al-sihhah* (hygiene and public health), and food has played an important therapeutic role in the Islamic world to this day--just one example of the connection of Islamic science with society and Islamic institutions.

Conclusion

The purpose of Islamic science and the role of Muslim scientists has not been merely to hand over to Europe what they had acquired from the Greeks and other ancients. Rather, having mastered what they learned from their predecessors, they were able to enrich it by their methods and new techniques. Many studies by Muslim scientists were transmitted to Europe in the Middle Ages as part of a wave of translation from Arabic into Latin in the twelfth and thirteenth centuries.

Islamic science reached its highest stage of development between the ninth and eleventh centuries, and subsequently experienced a number of major revivals during the twelfth and thirteenth centuries. The translation into Latin of the major Islamic works revived the spirit of learning in Western Europe during the late Middle Ages. The works of great Muslim authors such as ar-Razi, Ibn Sina and Ibn Rushd were widely read and frequently cited and quoted by Western writers.

All this flowed from the stimulus provided by the Quran which glorified Allah by wonder at His creation. Islam did not offer any kind of opposition to scientific research; in fact, it was quite the contrary. It is only when fetters are put on free research, in the name of official orthodoxy, and scientists are

Science and Muslim Societies

subjected to persecution and confinement that the edifice of science falls to the ground. Practising science within the ethical framework of Islam, Muslim scientists not only preserved the classical achievements of the ancients but also added new and original data to the reservoir of human knowledge.

In order to do full justice to the importance of their work, contemporary Western scientists must put into their historical context those who were, in former times, the teachers of their ancestors. Anawati, in Holt, et al, pp. 778-9

4

Islam and Science Education

Beliefs and Values in Science Education

> Pupils should develop their knowledge and understanding of the ways in which scientific ideas change through time and how the nature of these ideas and the uses to which they are put are affected by the social, moral, spiritual and cultural contexts in which they are developed.
> *The Nature of Science
> from the National Curriculum statutory orders for science
> preamble to attainment target 17*

Beliefs and values cannot easily be avoided in education, and neither should they be avoided. In fact, it is imperative that they be given central importance in the educational system because we all base our views and opinions on certain fundamentally held or vaguely perceived beliefs and values.

Recent British government legislation has restated the educational importance of beliefs and values. The 1988 British Educational Reform Act requires that a maintained school shall have 'a balanced and broadly based curriculum which promotes the spiritual, moral, cultural, mental and physical development of pupils'. The spiritual and moral aspects are not exclusive to religious education, as the encouragement of cross-curricular links shows. The requirements of attainment target seventeen as summarized in the preamble above, amply demonstrate this. For a full list of the seventeen science attainment targets of the British National Curriculum, see the appendix.

Value-laden topics permeate science education. The use of the environment, modern forms of genetic engineering, nuclear power, sexuality and issues concerning the inception and termination of life, are examples of the range of topics encountered within science education and related subjects. The Education Reform Act requires agreed syllabuses for religious education to 'reflect the fact that the religious traditions in Great Britain are in the main Christian whilst taking account of the teaching and practices of the other principal religions represented in Great Britain'. Part of the fulfilment of this broad requirement is that carefully thought-out Christian perspectives should be developed where cross-disciplinary links occur. Science education is probably the most important area where such value-laden perspectives are being developed.

The National Curriculum permits the development of perspectives other than Christian. In fact, the Christian perspective is not even the main one. In a democratic education system where majority views prevail, the dominant view takes the central place. Today, this happens to be secular humanism, a philosophy reflected throughout the curriculum where little emphasis is given to religion outside religious education. The dichotomy between religion and state is plainly manifested in the educational system which, unfortunately, has been adopted by most Muslim countries.

Any religious perspectives on science education are, therefore, marginal to the central perspective, namely of secular humanism. Attainment target seventeen does, nevertheless, allow the development of an Islamic perspective on certain issues of science education.

In a secular education system, the treatment of value-laden issues in science education presents a number of difficulties. Whereas there is a degree of consensus among the practitioners of science about what constitutes science and what are the criteria for adjudicating between rival views, there is much less agreement about beliefs and values. Because there are no empirical tests to adjudicate between different moral judgements, science is often labelled as objective, and beliefs and values as subjective. This gives the implication, as is probably the intention, that beliefs and values are of little importance except to the individual, having little or no relevance to wider society, let alone humanity. Such labelling is, of course, simplistic, unwise and blatantly incorrect. In a post-Kuhnian world it is untenable to argue for the complete objectivity of scientific knowledge. Although objectivity may be the aim of modern science, a whole cluster of subjective factors affects perception, theory selection and the handling of data. Conversely, beliefs and values are not simply matters of personal preference but have a reality external to the individual in binding society with a moral fibre that strengthens its ethical base. A secular education system lacks this holistic attitude to life, giving scant attention to promoting the positive values of godliness and holiness.

Humanizing Science in the Classroom

According to the aptly labelled 'alternative frameworks' educational theory, young people come into the classroom with certain vague or more elaborate, though scientifically erroneous, ideas about various scientific concepts. These 'primitive' ideas are based upon their own experiences, in turn influenced by their homes and social environment. As far as teaching is concerned, it is claimed by educational theorists that it is better to work with the alternative framework conceptions of the young person, gradually supplanting them with the 'true' scientific conceptions, rather than laying down autocratically at every lesson that which is scientifically right and worthy of belief. Whatever the scientific concepts studied in the classroom, however, one thing is certain: the students know that the influence of science is extremely strong in our society.

The students are aware of the advances that have come from the study and application of scientific principles. Just consider, as one example, the number of people on our planet who can witness simultaneously a person space-walking as it actually happens, thanks to the sophistication of modern telecommunications. So inexorable has been the march of scientific progress that many students come to believe that any material problem, be it a cure for cancer or an alternative fuel that makes an insignificant contribution to the greenhouse effect, can eventually be solved. This type of blind faith in science has found its way into the classroom from an outside world where success, prestige and worth are judged in materialistic terms, and where human progress is unashamedly equated with material progress. It is little wonder that students develop an intellect which assumes that science can solve our balance of payments deficit, improve our standard of living and give us all better health as well as a more desirable environment.

That such a philosophy can go seriously wrong is rarely pointed out in the classroom. Mrs Gandhi, when Prime Minister of India, found this out to her political cost, when her government embarked upon a ruthless programme of mass male sterilization to counter problems of over-population. This scientific solution had completely ignored and steamrollered over the profound religious and cultural beliefs that existed among the so-called uneducated sections of the rural population. It became clear that the scientific solution was no solution at all to a social problem. Examples like this serve to reinforce a fact which is really quite evident, namely, that no single facet of knowledge such as science can give us a wholly satisfying interpretation of our existence.

Science in the classroom often conveys the impression

that it is an impersonal enterprise. While it certainly needs people, somehow it functions independently of them. When I talk to students about the nature of science, the general message I get is that personal opinions and value judgements seemingly have no part to play in science. Science, therefore, appears to them as more authoritative and certainly more reliable than such activities as historical analysis, political discussion and moral debate where human opinions and prejudices are all too clearly apparent. Such a view, in my opinion, is based on a dangerous misunderstanding of science and scientific methods which science teachers themselves in their naive inductivism have inadvertently perpetrated, to give them the benefit of the doubt.

The human dimension must become integral to science teaching, for the process of scientific discovery is not an impersonal one: the human dimension is, to the contrary, of central importance. At some point in the formulation of a scientific hypothesis there is a step which lies beyond logical, rational explanation, which serves to produce something that is more than just the sum of the observable, experimental facts. Where relevant, examples from the history of science should be used in the classroom to reinforce this point. For example, the human dimension in scientific discovery is much in evidence in the dramatic account of the unravelling of the DNA structure in James Watson's book *The Double Helix*, and in Patrick Steptoe's and Robert Edwards's account of the events that led to the birth of Louise Brown, the first so-called test-tube baby.

That is precisely why there is so much important interaction between scientists and society. Society is usually ready to accept the benefits of science but often finds it difficult to live with the dangers that go with them. We should not wonder, therefore, that tensions exist between the scientist and society, often characterized in terms of a contrast between push and pull. Ideas and products generated by research push outwards from the laboratory bench into society, causing ripples and reverberations. At other times, society exerts a pull over the scientific enterprise, thereby shaping the work scientists and their institutions do. Social, political and commercial pressures within the scientific community also have a significant influence in the interaction with society.

To humanize science in the classroom, therefore, science must be taught to students in a manner that puts it into an appropriate social and cultural context. The applications and implications of scientific discoveries must be a central theme of science courses, not just optional extras. The British National Curriculum for science has now (at long last) gone some way towards meeting these essential requirements.

The Nature of Science, the rubric at the head of attainment target seventeen (reproduced at the beginning of this chapter) gives importance to the social, moral, spiritual and cultural contexts in which scientific ideas and theories develop. The term 'spiritual' again receives specific mention in one of the statements in attainment target seventeen, declaring that students should:

> Be able to give an historical account of a change in accepted theory or explanation, and demonstrate an understanding of its effects on people's lives--physically, socially, spiritually and morally, for example, understanding the ecological balance and the greater concern for our environment; the observations of the motion of Jupiter's moons and Galileo's dispute with the Church.

The examples given in the document are only illustrative and others, such as the Darwinian controversies, would clearly fit the requirements. Another section of the document stipulates that students should:

> Distinguish between claims and arguments based on scientific considerations and those which are not. Consider how the

development of a particular scientific idea or theory relates to its historical, and cultural--including the spiritual and moral--context. Study examples of scientific controversies and the ways in which scientific ideas have changed.

Again, in connection with humanizing science in the classroom, there are places within British science education and religious education for treating issues of science and religion. Because science and religion have been separated and compartmentalized, we do not expect them to be taught in a comprehensive and integrated fashion. Recent legislation has, nevertheless, opened up a new place for teaching about science *and* religion within an assessed part of science in the British National Curriculum.

Science and Religion: Tackling Problematic Issues

As a result of a monumental historical blunder, the relationship between science and religion is quite often misunderstood by students, with many actually believing that the two are incompatible. The points of confusion most commonly found can be expressed in the following eight statements which contain views quite often expressed by British students (usually non-Muslim, but a significant number of Muslims also).

1. If God exists, you should be able to prove it scientifically.
2. Man is nothing but a highly complex chemical mechanism.
3. Both scientific and religious explanations of the same events cannot be accepted.
4. If life comes from God, scientists will never be able to discover how it arises.
5. The statements 'God made Man' and 'Man was the result of an evolutionary process' need not be contradictory.
6. Religious belief can be explained in psychological terms.

7. The nature of scientific laws makes miracles an impossibilty.
8. Faith plays no part in science.

1. *If God exists, you should be able to prove it scientifically.*
The idea that everything of importance can be 'proved scientifically' is as erroneous as it is common. Modern science is not concerned with ultimate causes and is therefore silent about God. Meaning, purpose and plan in the universe are not issues addressed by science, and neither do the various methods of science provide the appropriate means for answering these questions, the answers to which must be sought elsewhere than in science.

> Not for (idle) sport did We create the heavens and the earth and all that is between.   Quran 21:16

2. *Man is nothing but a highly complex chemical mechanism.*
Although it is true that man is a highly complex chemical mechanism, when the words 'nothing but' are added the sentence becomes incorrect. Human beings are obviously much more than the chemicals of which they are composed. A solely chemical explanation will not say anything about whether a person is honest, compassionate, trustworthy or an atheist. It would be of limited use in getting to know somebody on a personal level. It is therefore a mistake to think that the chemical description of a person is the only one worth having, except for certain limited puroposes. To reduce descriptions solely to the level of atoms and molecules is to commit the error of reductionism. Here, holistic perspectives are sacrificed on the altar of parochial scientific models.

3. *Both scientific and religious explanations of the same events cannot be accepted.*
It is a well-known fact that there is more than one way of describing an object or event. Human knowledge provides several distinct ways of interpreting the world around us and

Islam and Science Education

the particular study which we call science is only one of these. Most significant social issues make demands on several different kinds of understanding. The routing of a new trunk road, for example, will require careful mathematical analyses of traffic patterns, scientific studies of soil structures, and moral and aesthetic arguments about the effect of the road on the quality of the environment.

Such different kinds of understanding and description together add up to a more complete picture of the whole. They are therefore complementary accounts. The scientific account of an object or event on its own plays only one part in the interpretation of our world and our relationship with it; it is never the whole, and neither could it be. The problem is that the deification of science has become a worldwide disease, reflected in the education system, science apparently outweighing other experiences in people's lives.

It is obvious that scientific accounts of objects and events, however complete in terms of mass, volume and chemical composition, say nothing about their origin or purpose. Religion, however, is very much concerned with both author and purpose, giving a comprehensive viewpoint which encompasses both the mechanism and the meaning.

> Behold! In the creation of the heavens and the earth; in the alternation of the night and the day. In the sailing of the ships through the ocean for the profit of mankind; in the rain which God sends down from the skies, and the life which He gives therewith to an earth that is dead. In the beasts of all kinds that He scatters through the earth. In the change of the winds, and the clouds which they trail like their slaves between the sky and the earth. Here indeed are signs for a people that are wise.
>
> Quran 2:164

4. *If life comes from God, scientists will never be able to discover how it arises.*

For a long time the idea that God is responsible only for those things which as yet have no scientific explanation and are labelled as 'mysterious', has been prevalent in the Western world. Such 'god-of-the-gaps' type of thinking ascribed thunder and lightening to the works of God until science explained it in terms of the flow of electrons causing very rapid heating of the air. This explanation disturbed Western theologians greatly, because they felt that the increasing success of science in explicating natural processes was a grave threat to biblical belief in God.

In the Islamic world such a mistake was never made. It was clear from the outset that the belief that life comes from God does not in any way imply that scientists cannot discover the processes involved. It is a fallacy to foster the misleading impression that the more scientific discoveries are made, the more God is pushed out of the universal picture. The discovery of complex biochemical processes which underlie life should provide fresh sources of wonder and further reason for worship rather than create doubts concerning God's existence. This type of spirit, it is worth remembering, provided a powerful stimulus to the work of many of the early Muslim scientists. They found encouragement in their studies from the conviction that they were uncovering the marvels of God's creation. Students should be made aware of this fact.

5. *The statements 'God made Man' and 'Man was the result of an evolutionary process' need not be contradictory.*

Evolution is a process, and a theoretical proposition; natural selection is the mechanism for that process. This statement is logically distinct from, and in no way interferes with, the statement that God created man. This is not to say, of course, that man arose as a result of an evolutionary process directed by God rather than by instantaneous creation. It just means that evolution is the name of a process and is in no way an ultimate cause. This type of misconception is quite prevalent among students and results in doubts (sometimes quite serious) about the existence of God. In fact, for hundreds of years it has been a common mistake in the Western world to think that the discovery of the way in which the universe functions somehow negates the idea that God is behind it. Some modern Muslims also hold such an illogical and erroneous view. It needs to be stressed that because the universe functions according to certain regular patterns, to conclude that it therefore functions by itself is an illogical corollary.

6. *Religious belief can be explained in psychological terms [or sociological or anthropological].*
That religious belief can be explained in psychological terms is not in doubt. Psychologists may study religious experience as one aspect of behaviour and try to explain it in psychological language, using the standard intellectual jargon. What is unacceptable, however, is to go a step further to twist the argument by claiming that modern psychology has explained away religious belief. To equate 'explaining' with 'explaining away' is the height of illogicality. An explanation of religious belief in psychological terms says nothing about the truth or otherwise of the beliefs held, which must be decided on other grounds, for example, by examining religious scriptures, historical evidence and other primary and secondary sources.

The same point needs to be made about sociological and anthropological studies of religion. Such accounts may look at the ways in which religious belief stabilizes a community or legitimates authority, that is, the function that religion plays in society. It is wholly inaccurate, however, to claim that such studies explain away religion just because they have a theoretical explanation for the role and purpose of religion in society. The truth of religious claims must be assessed on the grounds suggested above.

> Have they not then pondered over the Quran. If it had been from other than Allah, surely they would have found therein much inconsistency, much contracdiction.   Quran 4:82

7. *The nature of scientific laws makes miracles an impossibility.*

Science shows that nature is, to a large extent, orderly in its operation. This is why, it could be pointed out, its behaviour can be described in terms of certain simple laws. The main confusion which arises, however, springs from the incorrect belief that scientific laws are prescriptive, that is, they tell us what should take place rather than what does take place. It needs to be stressed that scientific laws are not commands to nature as to how it should behave, but concise, manmade descriptions of the ways in which it normally does behave. The fact that most behaviour is susceptible to such description does not exclude the possibility of exceptions.

On the basis of our previous experience, scientific laws may indicate what is expected to occur or what is unlikely to happen. They do not, however, make pronouncements as to whether something outside our present understanding can or cannot happen. Probability rather than possibility is the operative word for scientific method. That is to say, there is nothing absolute about manmade scientific laws, which are provisional and subject to modification whenever they clash with empirical analysis.

From the point of view of monotheistic faith, God is held to be the Creator of the world and, therefore, He formulated the laws. As their originator, it is presumably in His power to go against them if He so wills. Because He is omniscient and omnipotent, then, such occasions do not even need to be regarded as unnatural 'violations' of the laws. Science is accustomed to the idea that its laws have only limited domains of applicability. Newton's laws of motion, for instance, are perfectly fine for everyday purposes involving macroscopic physics, even for getting the Voyager spacecraft to Neptune. However, when it comes to handling situations involving infinitesimal particles that move at incredibly high velocities, then Newtonian mechanics reaches its limit and Einsteinian relativity and quantum physics come into play. In such cases, we do not say that Newton's laws are violated, rather, that these situations lie beyond the domain of applicability of Newton's laws. An analogous argument can therefore be put forward in connection with miraculous events that occurred during the ministries of various Prophets of God. If these unusual events--such as giving sight to a blind man, or bringing a dead person to life or reciting revealed verses that cannot be matched by the collective intellect of human beings--cannot be understood in terms of the normal operation of the laws of physics, then so be it. It is not that the physical law has been 'violated', rather, the event in question lies beyond its domain of applicability.

> If all mankind and jinn were to gather together to produce the like of this Quran, they would never be able to do so, even if they backed each other with help and support.   Quran 41:53

8. *Faith plays no part in science.*
Faith involves moving into the realm of the unknown on the basis of the known, on varying degrees of evidence. One may, for instance, have faith that a particular medicine will gradually eliminate an ailment if sufficient evidence shows that other people with the same illness have been cured by this medicine. The evidence, however, can be flimsy or even inaccurate. In fact, too many people are too ready to believe anything on insufficient evidence: this is when faith is more accurately termed credulity. The evidence on which we base our faith is therefore all-important, sincerity of belief alone not being enough because it is quite possible to be sincerely wrong.

Those who think that faith plays no part in science are, in fact, sincerely wrong. Consider, for example, the process called 'extrapolation', familiar to most students of science and sometimes used when drawing a graph. Points which have been obtained experimentally by testing the relationship between two variables are joined by a line. Sometimes it is necessary and informative to extend the line into unknown regions where no points have been obtained. Underlying such extrapolation is the implicit assumption that similar behaviour occurs in the unexplored part of the experiment as in the explored part, and that the results in the unknown region can be predicted on the strength of the known. This is surely an expression of faith in the empirical data one has obtained.

Faith, then, is also involved in science as it is in other areas of life, In fact, before science can even begin the scientist must have faith in the comprehensibility of the universe, that the universe is intelligible and capable of being understood. Any scientist who did not believe this would not begin to scratch the surface of natural phenomena, let alone plumb the depths of the physical and biological behaviour of the natural world.

If scientists believe that the universe is intellectually comprehensible, then they have faith in the power of human reason to make valid deductions from empirical data. That is to say, they have faith in the rationality of the human mind. Unfortunately, however, a substantial number of non-Muslim scientists still reject the view that we have been created according to a divine plan and with a purpose. Consider the following puzzle which many of those who reject God have difficulty in explaining:

> If I am the result of an unplanned collection of atoms, then my brain as part of me is also an unplanned collection of atoms. However, if my brain is an unplanned collection of atoms, how can I expect that the conclusions it comes to are true ones? How can I then know that I am the result of an unplanned collection of atoms?

The Quran says majestically:

> So set your purpose (O Mankind) towards religion as men by

nature upright. The nature framed of Allah in which He has created man. There is no altering the laws of Allah's creation. That is the right religion; but most of mankind know not.

Quran 30:30

Behind the search for patterns of behaviour in the natural world, then, lies the belief that the universe is essentially orderly, and that the human mind finds it intelligible. Such patterns are described concisely in the form of scientific laws--intellectual regulations without which there would be no science. The belief that orderliness is to be found throughout nature is in itself an act of faith.

At one time in the Western world, such acts of faith arose directly out of the scientist's biblical belief in God as the Creator of the universe. Deism, however, then took over, seventeenth-century mechanicism espousing the idea of a clockwork universe in which a once active God had the master key, but now stood passively by to see the results of His work unfold as the universal clock ticked with astonishingly predictable regularity. Agnosticism, with its fraternal twin atheism, was not far off as God was gradually pushed out of man's life and into the mythological wilderness.

We now have the unedifying situation in which belief in the orderliness and intelligibility of the universe acts as a stimulus to persevere with the investigation of a difficult problem but coupled with a low degree of belief in God, primarily in the Western scientific world. From the Islamic point of view, of course, it is because God has indicated in the Quran that He is a God of order and not of chaos that the scientist may expect to find the universe to be both orderly and intelligible.

We [God] will show them the signs in all regions of the earth and in themselves until they come to see that this is the truth.

Quran 41:53

Science and Muslim Societies

Integrating Religion and Science in the Classroom

The modern secular education system has a profound influence on the thinking of adolescents, which has led to a largely unquestioned acceptance of science as a worldview and the rejection of the importance of religion, literally from the sublime to the ridiculous. The influence of secular metaphors is unintentionally, and occasionally intentionally, reinforced by science teaching, particularly true if science is taught in a vacuum as though it has no religious context.

During my teaching practice a few years ago, I had the experience of teaching both biology and religious education in a London secondary school. It came as a rude surprise to me that my credibility with the students varied depending on which subject I was teaching. In biology, students accepted whatever I taught without question, voraciously consuming all given facts. In religious education, however, I was challenged on virtually everything I said. When I told them that for Muslims the Quran is the literal word of God and that I personally believed this to be a fact, I was accused of 'shoving religion down their throats'. The response was astonishingly different, however, when I told them that according to the evolutionary theory life arose from non-life by chance events. For the average secondary school student in the Western world, to doubt in science is often seen as irrational as it is for them to believe in religion.

In the minds of most Western adolescents there is an exaggerated emphasis on the divide between religious and scientific ways of perceiving the world. The religious is often perceived as being 'illusory', whereas the scientific is 'real and true'. Ironically, however, although they have largely rejected institutional and traditional religion, there is a marked interest in the supernatural, particularly where there is a science-fiction element in it. Students often accept second or even third-hand evidence for flying saucers and UFOs but impugn the evidence for the existence of God.

Islam and Science Education

Scientific evidence is equated with acceptable evidence. Is it surprising, then, when they conclude that it is in the laboratory and not in the mosque, temple or church where the future of humanity lies. Muslim students in the Western world (and I dare say in the Muslim world, too) are vulnerable to such naive intellectual influences. The problem is real and cannot conveniently be swept under the carpet.

One of the big problems is that many science teachers are out of touch with developments in the philosophy of science. They operate on a Baconian, inductivist view of scientific method which sees scientific knowledge as based on certain verifiable suppositions. This supports the popular, commonsense notion of science as certain, based on uncontroversial, totally objective observation. The idea that science is a superior form of knowing is reinforced and a clear distinction is made between science lessons that deal with facts and religious education lessons that deal with mere opinions.

From the point of view of Islam, it is extremely important that science teaching does not perpetuate the misleading 'popular' view of science, since this reinforces the Western dualism between scientific and religious modes of thinking. In the absence of carefully planned teaching, students will automatically adopt the prevalent secular metaphors and accept (incorrectly, of course) that there is a conflict between science and religion. It is important to grasp this point. So powerful are the secular metaphors in the modern-day educational system that if the science teacher in a lesson about evolution explains the process of natural selection but refuses to discuss the notion of creation then most students will automatically assume that evolution has displaced creation. This is a fundamental problem of a society in which a sharp distinction is maintained between the affairs of religion and the affairs of the state.

The task of science teachers, therefore, is vitally important. They have a moral duty to help students to integrate their scientific knowledge into an overall understanding of humans, life and the universe. If they maintain autonomy between the scientific and religious domains of thought they will, by default, reinforce the hidden messages of the secular metaphors and thereby promote the popular misconception of the exclusivity of scientific knowledge in the minds of adolescent students.

The idea that scientific and religious knowledge are independent of each other is a worldview that perpetuates a Western secular view of knowledge through its stress on the irrelevance of placing science in a religious context. Admittedly (and commendably), however, some positive progress has been made in this connection in the form of attainment target seventeen of the British national science curriculum. The nature of science, as it is called, does make some attempt, albeit minor, towards contextualizing scientific theories within religious and spiritual settings. The General Certificate of Secondary Education (GCSE) criterion for science also recognizes the importance of placing science in its social, economic, technological, ethical and cultural contexts. The limitations of science are at last being stressed, getting away from the idea of uncontroversial objectivity and incontrovertibility.

Although from the Islamic point of view these developments leave much to be desired, they should be commended as small but positively contributory steps in the right direction. The pluralism of worldviews on which the emerging British educational philosophy is built should not, however, be the accepted practice of the Muslim world. We have our own clear and distinctive worldview of humans, life and the universe--the solid foundation upon which the giant edifice of the Islamic educational system is erected. As Seyyed Hossein Nasr has said in a perceptive educational article,

> Modern science is based on a specific worldview. This science cannot be taught in the Muslim world as if the worldview on

Islam and Science Education

which it is based is not there, or as if Islam did not possess its own perspectives....Science education according to the Islamic perspective must begin and have as its background throughout all the stages of education, this quranic view of the cosmos.

Nasr, 'Science Education'

God is He, than Whom there is no other god; the Sovereign, the Holy One, the Source of Peace (and Perfection), the G... of Faith, the Preserver of Safety, the Exalted in Might, the Irresistible, the Supreme.

Glory to God! (High is He) above the partners they attibute to Him.

He is God, the Creator, the Evolver, the Bestower of Forms (or Colours). To Him belong the Most Beautiful Names. Whatever is in the heavens and on earth, does declare His Praises and Glory; and He is the Exalted in Might, the Wise.

Quran 49:22-24

Bibliography

Ahmad, N. *Muslim Contribution to Geography*. Lahore: Ashraf, 1965.

Al-Daffa, A.A. *The Muslim Contribution to Mathematics*. London: Croom Helm, 1977.

Anees, Munawar Ahmad. 'Islamic Science--An Antidote to Reductionism', *Inquiry* I(2):49 (July 1984).

_____. 'What Islamic Science Is Not', *MAAS Journal of Islamic Science* 2(1):9-20 (January 1986).

_____. *Islam and Biological Futures: Ethics, Gender and Technology*. London and New York: Mansell, 1989.

Bacon, F. (ed. by F. Anderson). *The New Organon*. Indianapolis: Bobbs-Merrill, 1975 (first published 1620).

Bernal. J.D. *The Social Function of Science*. Cambridge, Massachusetts: MIT Press, 1967 (first published 1939).

Bucaille, Maurice. *The Bible, the Quran and Science*. Paris: Seghers, 1976.

Chalmers, A.F. *What Is This Thing Called Science?* Oxford: OUP, 1982.

Chughtai, M.I.D. 'Role of Scientific Societies in the Promotion of Science in the Islamic World', *Science, Technology and Development* 2(1):2-3 (1983).

Department of Education and Science. *Science in the National Curriculum.* London: HMSO, 1989.

Dictionary of Scientific Biography. New York: Scribners, 1981.

Drake, S. *Galileo at Work: His Scientific Biography.* Chicago, UCP, 1978.

Dunlop, D.M. *Arab Civilization to AD 1500.* London: Longman, 1971.

Durrani, S.A. 'Importance of Scientific Research in Islamic Countries: A Blueprint for Progress', *MAAS Journal of Islamic Science* 2(2):23 (1986).

Easler, Brian. *Liberation and the Aims of Science: An Essay on the Obstacles to the Building of a World.* London: Chatto and Windus, 1973.

El-Nejjar, Z.R. 'The Limitations of Science and the Teachings of Science from the Islamic Perspective', *American Journal of Islamic Social Sciences* 3(1):59 (1986).

Feyerabend, Paul. *Against Method: Outline of an Anarchistic Theory of Knowledge.* London: New Left Books, 1975.

_____. *Science in a Free Society.* London: New Left Books, 1978.

Freudenthal, Gideon. *Atom and Individual in the Age of Newton*. Reidel: Dordrecht, 1986.

Graham, L. *Between Science and Values*. New York: Columbia University Press, 1981.

Greenberg, D.S. *The Politics of Pure Science*. New American Library, 1971.

Habermas, J. *Knowledge and Human Interests*. London: Heinemann, 1972.

Haq, S. 'The Quran and Modern Cosmologies', *Science and Technology in the Islamic World* **1**:47-52 (1983).

Hashmi, Z.A. 'Science in the Islamic World', *Science Technology and Development* **1**(1):2-7 (1982).

Hirst, P. *Moral Education in a Secular Society*. University of London Press, 1974.

Holt, P.M., A. Lambton and B. Lewis (eds.). *The Cambridge History of Islam*. Vol. 2B. Cambridge: CUP, 1970.

Jeffreys, A. *Scientific Inference*. Oxford: OUP, 1931.

Kettani, M.A. 'Science and Technology and the Muslim World', *MAAS Journal of Islamic Science* **2**(2):49 (1986).

Kibble, Parker and Price. 'The Age of Uncertainty--Religious Beliefs Among Adolescents', *British Journal of Religious Education* **4**(1):31-5 (1981).

Kuhn, T.S. *The Structure of Scientific Revolutions*. Chicago: UCP, 1962.

Lakatos, I. and A. Musgrave (eds.). *Criticiswm and the Growth of Knowledge*. Cambridge, CUP, 1974.

Manzoor, S. Parvez. 'The Thinking Artifice: AI and Its Discontents', *Inquiry* 3(9):34-9 (1986).

Malthus, Thomas R. *An Essay on the Principle of Population*. Harmondsworth: Penguin, 1970.

Martin, B. and R. Pluck. *Young People's Beliefs*. London: General Synod Board of Education, 1977.

Maxwell, Nicholas. 'Science, Reason, Knowledge and Wisdom: A Critique of Specialism', *Inquiry* 23:19-81 (1980).

_____. *From Knowledge to Wisdom: A Revolution in the Aims and Methods of Science*. Oxford: Basil Blackwell, 1984.

Michael, Donald. 'Competence and Compassion in an Age of Uncertainty', *World Future Society Bulletin* (Jan-Feb 1983).

Midgeley, M. *Evolution as a Religion*. London: Methuen, 1985.

Mitroff, Ian. *The Subjective Side of Science*. Amsterdam: Elsevier, 1974.

Nasr, Seyyed Hossein. *The Encounter of Man and Nature*. Cambridge, Massachusetts: Harvard University Press, 1968.

_____. *Science and Civilization in Islam*. Cambridge, Massachusetts: Harvard University Press, 1968.

_____. *Islamic Science: An Illustrated Study*. London: World of Islam Festival, 1976.

_____. 'Science Education: The Islamic Perspective', *Muslim*

Education Quarterly **5**(1):4-14 (1987).

Nasseef, A.O. 'The Role of Faith and Islamic Ethics in the Teachings of Natural Sciences', *Muslim Educational Quarterly* **1**(2):7 (1982).

———. 'Science, Education and Religious Values: An Islamic Approach', *Muslim Education Quarterly* **1**(3):4 (1984).

Newton-Smith, W.H. *The Rationality of Science*. London: Routledge & Kegan Paul, 1982.

Peacocke, A.R. *Creation and the World of Science*. Oxford: Clarendon Press, 1988.

Polanyi, Michael. *Personal Knowledge*. London: Routledge & Kegan Paul, 1958.

Poole, M.W. 'An Investigation into Aspects of the Interplay Between Science and Religion at Sixth Form Level'. MPhil thesis, Kings College, University of London, 1983.

———. 'Science and Education and the Interplay Between Science and Religion', *Schools Science Review* **67**:239-44 (1989).

Popper, Karl R. *The Logic of Scientific Discovery*. London: Hutchinson, 1959 (first published 1934).

———. *Conjectures and Refutations*. London: Routledge & Kegan Paul, 1963.

———. *The Open Society and Its Enemies*. London: Routledge & Kegan Paul, 1969 (first published 1945).

———. *Objective Knowledge*. Oxford: OUP, 1972.

Ravetz, Jerome R. *Scientific Knowledge and its Social Problems*. Oxford: OUP, 1971.

_____. *The Merger of Knowledge with Power: Essays in Critical Science*. London and New York: Mansell, 1990.

Reich, H. 'Between Religion and Science: Complementarity in the Religious Thinking of Young People, *British Journal of Religious Education* 11(2), 62-9, 1989.

Rose, Hilary and Stephen Rose (eds.). *Ideology of/in the Natural Sciences*. London: Macmillan, 1976 (2 vols.).

Rossi, P. *Philosophy, Technology and the Arts in the Early Modern Era*. New York, Harper Torchbrook, 1970.

Roszak, Theodore. *The Making of a Counter Culture*. London: Faber, 1970.

Russell, Bertrand. *A History of Western Philosophy*. London: Allen & Unwin, 1979 (first published 1946.

Salam, M.A. 'Islam and Science', *MAAS Journal of Islamic Science* 2(1):21 (1986).

Sardar, Ziauddin. *Arguments for Islamic Science*. Aligarh: Centre for Studies on Science, 1985.

_____. *Islamic Futures: The Shape of Ideas to Come*. London and New York: Mansell, 1985.

_____. *Information and the Muslim World: A Strategy for the Twenty-First Century*. London and New York: Mansell, 1988.

_____. *Explorations in Islamic Science*. London and New York: Mansell, 1989.

_____ (ed.). *The Touch of Midas: Science, Values and Environment in Islam and the West*. Manchester: MUP, 1984.

_____ (ed.). *The Revenge of Athena: Science, Exploitation and the Third World*. London and New York: Mansell, 1988.

_____ (ed.). *An Early Crescent: The Future of Knowledge and the Environment in Islam*. London and New York: Mansell, 1989.

Sarton, G. *Introduction to the History of Science* (3 vols.). Baltimore: Williams and Wilkins, 1927.

Ullmann, M. *Islamic Medicine*. Edinburgh: EUP, 1978.

Watson, B. *Educational Belief*. Oxford: Basil Blackwell, 1987.

Wickens, G. (ed.). *Avicennà (Ibn Sina): Scientist and Philosopher*. London: Luzac, 1952.

Young, Robert M. 'Science *is* Social Relations', *Radical Science Journal* **9**:61-131 (1977).

Ziman, J. *The Force of Knowledge: The Scientific Dimension of Society*. Cambridge: CUP, 1976.

Appendix 1

Science Attainment Targets of the British National Curriculum (1989)

Attainment targets are defined as:

> The knowledge and understanding which students of different maturities and abilities are expected to achieve at the end of each key stage.

Exploration of science, communication and the application of knowledge and understanding

    AT1   Exploration of Science

Knowledge and understanding of science, communication, and the applications and implications of science

- AT2 The variety of life
- AT3 Processes of life
- AT4 Genetics and evolution
- AT5 Human influences on the Earth
- AT6 Types and uses of materials
- AT7 Making new materials
- AT8 Explaining how materials behave
- AT9 Earth and atmosphere
- AT10 Forces

AT11	Electricity and magnetism
AT12	The scientific aspects of information technology including micro-electronics
AT13	Energy
AT14	Sound and music
AT15	Using light and electromagnetic radiation
AT16	The Earth in space
AT17	The nature of science

Attainment target 17: The nature of science

Pupils should develop their knowledge and understanding of the ways in which scientific ideas change through time and how the nature of these ideas and the uses to which they are put are affected by the social, moral, spiritual and cultural contexts in which they are developed; in doing so, they should begin to recognize that while science is an important way of thinking about experience, it is not the only way.

Level *Statement of Attainment*

Pupils should

4 Be able to give an account of some scientific advance, i.e., in the context of medicine, agriculture, industry or engineering, describing the new ideas and investigation or invention and the life and times of the principal scientist involved.

5 Be able to disuss clearly with others their way of thinking about some experiment which is new to them. Be able to demonstrate that different interpretations of the experimental evidence that they have collected are possible.

Appendix 1: Science Attainment Targets

6 Be able to use one or two explanatory models from their own learning in science to demonstrate how predictions have been made which stimulate new experiments.

Be able to describe and explain one incident from the history of science where successful predictions were made to establish a new model, i.e., the work of scientists on air-borne organisms (Pasteur) and the evidence for atmospheric pressure (Pascal).

7 Be able to give an historical account of a change in accepted theory or explanation, and demonstrate an understanding of its effects on people's lives: physically, socially, spiritually and morally, i.e., understanding the ecological balance and the greater concern for our environment, the observations of the motion of Jupiter's moons and Galileo's dispute with the Church.

Be able to demonstrate an appreciation of differing functions of scientific evidence and imaginative thought in carrying forward scientific understanding, i.e., discovery of the structure of DNA--the different approach of Franklin from that of Watson and Crick.

8 Be able to explain how a scientific explanation from a different culture or a different time contributes to our present understanding.

Understand the uses of evidence and the tentative nature of proof.

9 Be able to distinguish between generalizations and predictive theories and give an example of each, i.e., such pairs might be 'all metals conduct electricity' and 'the theory of a free electron gas which predicts this property', or 'a clear sky in winter always means frost at night' and 'the absence of clouds to reflect back the Earth's radiation is the basis of such a prediction'.

10 Be able to demonstrate an understanding of the differences in scientific opinion on some topic, either from the past or present, drawn from studying the relevant literature, i.e., plate tectonics and the wrinkling of a shrinking Earth or living things reproduce their own kind and the spontaneous generation of species.
Be able to relate differences of scientific opinion to the uncertain nature of scientific evidence, i.e., what is the cause of 'cot deaths?' or what is responsible for the death of trees in European forests?

N.B. The Department of Education and Science (DES) has now produced a heavily truncated version of the science statutory curriculum, reducing the seventeen attainment targets to five, designed to be implemented in September 1992. The intention is to reduce the onerous load of national curriculum teaching requirements on individual teachers in particular and on schools in general. The five new attainment targets for science are:

1. Scientific investigation
2. Life and living processes
3. Earth and the atmosphere
4. Materials and their uses
5. Energy and its effects

The content and thrust behind attainment target seventeen, The Nature of Science, is largely unchanged, however. Much of it will underpin attainment targets two to five. Together with attainment target one, its requirements will bind together the contents of the science statutory curriculum.

Appendix 2

Muslim Scientists

1. Abbas ibn Firnas (b. Ronda, Spain; d. 887). Humanities, technology.

Al-Abbas ibn Said al-Jawhari, *see* al-Jawhari.

2. Ibn Abi-Usaybiah, Muwaffaq-al-Din Abu-al-Abbas Ahmad (b. 1203 Damascus; d. 1270). History of Islamic medicine.

3. Abraham bar Hiyya ha-Nasi (fl. Barcelona before 1136). Mathematics, astronomy.

4. Abu Hamid al-Gharnati (b. Granada, 1080; d. Damascus, 1169). Geography.

5. Abu Kamil Shuja ibn Aslam ibn Muhammad ibn Shuja (c.850-c.930). Mathematics.

6. Abul-Barakat al-Baghdadi, Hibat Allah (b. Iraq, c.1080; d. Baghdad, c.1164-5). Physics, psychology, philosophy.

7. Abul-Fida Ismal ibn (b. Damascus, 1273; d. Hama, Syria, 1331). History, geography.

8 Abul Kasim, Abu-al-qasim Khalaf ibn-Abbas al-Zahrawi (b.

Cordova, c. 936; d. c.1013). Medicine, surgery.

9. Abul Wafa al-Buzjani, Muhammad ibn Muhammad ibn Yahya ibn Ismail ibn al-Abbas (b. Buzjan [now Iran], 940; d. Baghdad, 997-8). Astronomy, mathematics. One of the greatest Muslim mathematicians.

10. Abu Mashar al-Balkhi, Jafar ibn Muhammad (b. Khurasan, 787; d. Al-Wasit, Iraq, 886). Astrology.

11. Ibn Aflah al-Ishbili, Abu Muhammad Jabir (fl. Seville, first half 12th century). Astronomy, mathematics.

12. Al-Baghdadi, Abu Mansur Abd al-Qahir ibn Tahir ibn Muhammad ibn Abd Allah, al-Tamimi, al-Shafii (b. Baghdad; d. 1037). Arithmetic.

13. Ibn Bajja, Abu Bakr Muhammad ibn Yahya ibn al-Saigh (b. Saragossa, Spain; d. Fez, 1139). Philosophy.

14. al-Bakri, Abu Ubayd Abdallah ibn Abd al-Aziz ibn Muhammad (1040-94). Geography.

15. Ibn al-Banna al-Marrakushi (b. Marrakesh, 1256; d. Marrakesh, 1321). Mathematics.

16. Banu Musa [three brothers Muhammad, Ahmad and al-Hasan were known under the one name which means 'sons of Musa'] (b. Baghdad; fl.850; d. Baghdad). Mathematics, astronomy.

17. al-Battani, Abu Abd Allah Muhammad ibn Jabir ibn Sinan al-Raqqi al-Harrani al-Sabi [Latin, Albategnius or Albatenius] (b. near Harran [now Turkey], c.858; d. near Samarra, Iraq, 929). Astronomy, mathematics. The 'greatest Islamic astronomer'.

18. Ibn al-Baytar al-Malaqi, Diya al-Din Abu Muhammad Abd

Appendix 2: Muslim Scientists

Allah ibn Ahmad (b. Malaga, c.1190; d. Damascus, 1248). Pharmacology, botany.

19. al-Biruni, Abu Rayhan Muhammad ibn Ahmad (b. Khwarizm [now USSR], 973; d. Ghazna [now Ghazni, Afghanistan], 1048). Astronomy, chronology, mathematics, medicine, geography, history.

20. al-Bitruji al-Ishbili, Abu Ishaq [Latin, Alpetragius] (fl. Seville, c.1190). Astronomy, natural philosophy. Outstanding astronomer among the Spanish Aristotelians.

21. Ibn Butlan, Abul-Hasan al-Mukhtariclen Abdun ibn Sadun (b. Baghdad, c.1000; d. Antioch, 1068). Medicine.

22. al-Damiri, Muhammad ibn Musa (b. Damira, Egypt, 1344; d. Cairo, 1405). Natural history.

23. al-Dinawari, Abu Hanifah. Arithmetic, law.

24. Ibn Ezra, Abraham ben Meir, also known as Abu Ishaq al-Majid ibn Ezra (b. Toledo, c.1090; d. Calahorra, Spain, c.1164). Mathematics, astronomy.

25. al-Farabi, Abu Nasr Muhammad ibn Muhammad ibn Tarkhan ibn Awazlagh [Latin, Alfarabius] (b. Wasij, district of Farab in Central Asian province of Transoxiana, c.870; d. Damascus, 950). Philosophy, music.

26. al-Farghani, Abul-Abbas Ahmad ibn Muhammad ibn Kathir (b. Farghana, Transoxania; d. Egypt after 861). Astronomy.

27. al-Fazari, Muhammad ibn Ibrahim (fl. c.760-90). Astronomy.

28. al-Hamdani, Abu Muhammad al-Hasan ibn Ahmad ibn Yaqub, also known as Ibn al-Haik, Ibn Dhil-Dumayna, or Ibn Abil-

Dumayna (b. Sana, Yemen, c.893; d. Sana, 945). Geography, natural science.

29. Ibn Hawqal, Abul-Qasim Muhammad (b. Nisibis [now Nusaybin, Turkey]; fl. second half 10th century). Geography.

30. Ibn al-Haytham, Abu Ali al-Hasan ibn al-Hasan, called al-Basri, al-Misri (b. 965; d. Cairo, c.1040). Optics, astronomy, mathematics.

31. Ibn Hayyan, Jabir (b. Tus, Persia, c.738; d. Kufa, Iraq, c.813). Chemistry, alchemy.

32. al-Idrisi, Abu Abd Allah Muhammad ibn Muhammad ibn Abd Allah ibn Idris, al-Sharif (b. Centa, Morocco, 1100; d. Centa, 1166). Geography, cartography.

33. Ishaq ibn Hunayn, Abu Yaqub (d. Baghdad, c.910). Medicine, scientific translation.

34. Al-Jahiz, Abu Uthman Amr ibn Bahr (b. Basra, c.1766; d. Basra, 1869). Natural history.

35. al-Jawhari, al-Abbas ibn Said (fl. Baghdad, c.830). Mathematics, astronomy.

36. al-Jayyani, Abu Abd Allah Muhammad ibn Muadh (b. Cordoba, c.989; d. after 1079). Mathematics, astronomy.

37. al-Jazari, Badi al-Zaman Abul-Izz Ismail ibn al-Razzaz (fl. Diyarbakr, Turkey, 1206). Machinery, techniques of construction.

38. Ibn Jubair, Abu Husain Muhammad ibn Ahmad (b. Valencia, 1145; d. 1217). Geography.

39. Ibn Juljul, Sulayman ibn Hasan (b. Cordoba, 944; d. c.994).

Appendix 2: Muslim Scientists

Medicine, pharmacology.

40. Kamal al-Din Abul-Hasan Muhammad ibn al-Hasan al-Farisi (d. Tabriz, Iran, 1320). Optics, mathemicatics.

41. al-Karaji [or Karkhi] Abu Bakr ibn Muhammad ibn al-Husayn [or al-Hasan] (fl. Baghdad, c.1000). Mathematics.

42. al-Kashi [or Kashani], Ghiyath al-Din Jamshid Masud (b. Kashan, Iran; d. Samarkand [now in Uzbek, USSR], 1429). Astronomy, mathematics.

43. Ibn Khaldun [Abd-al-Rahman ibn Muhammad] (b. Tunis, 1332; d. Cairo, 1406). History, philosophy of history, sociology.

44. al-Khalili, Shams al-Din Abu Abd Allah Muhammad ibn Muhammad (fl. Damascus, c.1365). Astronomy, mathematics.

45. al-Khayyami [or Khayyam], Ghiyath al-Din Abul-Fath Umar ibn Ibrahim al-Nisaburi (or al-Naysaburi), also known as Omar Khayyam (b. Nishapur, Khurasan [now Iran], c.1048; d. Nishapur, c.1123). Mathematics (algebra), astronomy, philosophy.

46. al-Khazini, Abul-Fath Abd al-Rahman [sometimes Abu Mansur Abd al-Rahman or Abd al-Rahman Mansur] (fl. Merv [now Mary, Turkmen SSR, USSR], c.1115-30). Astronomy, mechanics, scientific instruments.

47. Al-Khujandi, Abu Mahmud Hamid ibn al-Khidr (d. 1000). Mathematics, astronomy.

48. Ibn Khurradadhbih [or Khurdadhbih], Abul-Qasim Ubayd Allah Abd Allah (c.820-c.912). Geography, history, music.

49. al-Khwarizmi, Abu Abd Allah Muhammad ibn Ahmad ibn Yuuf (fl. Khwarizm, 9th century). Mathematics, astronomy. His

work on elementary mathematics, *Kitab al jabr wal muqabala* has in part survived in the term 'algebra' and his name in the term 'algorism'.

50. al-Kindi, Abu Yusuf Yaqub ibn Ishaq al-Sabbah (b. c.801; d. Baghdad, c.866). Philosophy, science (astrology, medicine, Indian arithmetic), logogriphs. The first outstanding Islamic philosopher, called 'the philosopher of the Arabs'. The titles of more than 270 of his works are known.

51. Ibn Majid, Shihab al-Din Ahmad (fl. Najd, Saudi Arabia, 15th century). Navigation. No Muslim navigator of the Middle Ages surpassed Ibn Majid in the intimate knowledge and experience of both the Red Sea and the Indian Ocean.

52. al-Majriti, Abul-Qasim Maslamah ibn Ahmad ala-Faradi (b. Madrid; d. Cordoba, c.1007). Astronomy.

53. al-Majusi, Abul-Hasan Ali ibn Abbas [Haly Abbas] (b. al-Ahwaz-Khuzistan [near Shiraz, Iran]; d. Shiraz, 994). Medicine, pharmacology, natural science.

54. Mansur ibn Ali ibn Iraq, Abu Nasr (fl. Khwarizm; d. Ghazna, c.1036). Mathematics, astronomy.

55. al-Maqdisi [or Muqaddasi], Shams al-Din Abu Abd Allah Muhammad ibn Ahmad ibn Abi Bakr al-Banna al-Shami al Maqdisi al-Bashshari (b. Bayt al-Maqdis, Jerusalem, c.946; d. c.1000). Geography, cartography.

56. Masha Allah (fl. Baghdad, c.762-815). Astrology.

57. Ibn al-Nafis, Ala al-Din Abul-Hasan Ali ibn Abil-Hazm al-Qurashi (b. al-Qurashiyya, near Damascus; d. Cairo, 1288). Medicine.

Appendix 2: Muslim Scientists

58. al-Nasawi, Abul-Hasan, Ali ibn Ahmad (fl. Baghdad, 1029-44). Arithmetic, geometry.

59. al-Nayrizi, Abul-Abbas al-Fadi ibn Hatim (fl. Baghdad, c.897-922). Geometry, astronomy.

Omar Khayyam, *see* Khayyami

60. al-Qabisi, Abu al-Saqr Abd al-Aziz ibn Uthman ibn Ali (fl. Aleppo, Syria, c.950). Astrology.

61. Qadi Zada al-Rumi [Salah al-Din Musa Pasha] (b. Bursa, Turkey, c.1364; d. Samarkand, c.1436). Mathematics, astronomy.

62. al-Qalasadi [or Qalsadi], Abul-Hasan Ali ibn Muhammad ibn Ali (b. Basta [now Baza, Spain], 1412; d. Beja, Tunisia, 1486). Arithmetic, algebra, Islamic law.

63. al-Qazvini, Zakariyya ibn Muhammad ibn Mahmud, Abu Yahya (b. Qazvin [now Kasvin, Iran], c.1203; d. 1283). Cosmography, geography.

64. Ibn al-Quff, Amin al-Dawlah Abu al-Faraj ibn Muwaffaq al-Din Yaqub ibn Ishaq al-Masihi al-Karaki (b. Karak, Jordan, 1233; d. Damascus, 1286). Medicine, physiology, natural sciences, philosophy.

65. al-Quhi [or Kuhi], Abu Sahl Wayjan ibn Rustam (fl. Baghdad, c.970-1000). Mathematics, astronomy.

66. Qusta ibn Luqa al-Balabakki (fl. Baghdad and Armenia, 860-900). Medicine, philosophy, translation of scientific literature.

67. Ibn Qutayba, Abu Muhammad Abd Allah ibn Muslim al-Dinawari al-Jabali (b. Baghdad or Kufa, Iraq, 828; d. Baghdad, c.889). Transmission of knowledge.

68. Qutb al-Din al-Shirazi (b. Shiraz, Persia, 1236; d. Tabriz, Persia, 1311). Optics, astronomy, medicine, philosophy.

69. al-Razi, Abu Bakr Muhammad ibn Zakariyya [Rhazes] (b. Rayy, Persia, c.854; d. Rayy, c.935). Medicine, alchemy, philosophy, religious criticism.

70. Ibn Ridwan, Abul-Hasan Ali Ibn Ali ibn Jafar al-Misri (b. El-Ghiza, Egypt, 998; d. Cairo, c.1069). Medicine.

71. Ibn Rushd, Abul-Walid Muhammad ibn Ahmad ibn Muhammad [Averroes] (b. Cordoba, 1126; d. Marrakesh, 1198). Astronomy, philosophy, medicine.

72. al-Samarqandi, Najib al-Din Abu Hamid Muhammad ibn Ali ibn Umar (d. Herat, Afghanistan, 1222). Medicine, materia medica.

73. al-Samarqandi, Shams al-Din Muhammad ibn Ashraf al-Husayni (b. Samarkand, Vzbekistan, Russia, fl. 1276). Mathematics, logic, astronomy.

74. Ibn al-Shatir, Ala al-Din Abul Hasan Ali ibn Ibrahim (b. Damascus, c.1305; d. Damascus, c.1375). Astronomy.

75. al-Sijzi, Abu Said Ahmad ibn Muhammad ibn Abd al-Jalil (b. Sijistan, Persia, c.945; d. c.1020). Geometry, astronomy, astrology.

76. Ibn Sina, Abu Ali al-Husayn ibn Abd Allah [Avicenna] (b. Kharmaithen, near Bukhara, Persia [now USSR], 980; d. Hamadan, Persia, 1037). Philosophy, medicine, biology, astronomy.

77. Ibn Sinan, Ibrahim ibn Thabit ibn Qurrah (b. Baghdad, 908; d. Baghdad 946). Mathematics, astronomy.

78. Sinan ibn Thabit ibn Qurrah, Abu Said (b. c.880; d. Baghdad, 943). Medicine, astronomy, mathematics.

Appendix 2: Muslim Scientists

79. al-Sufi, Abul-Husayn Abd al-Rahman ibn Umar al-Razi (b. Rayy, Persia, 903; d. 986). Astronomy.

80. al-Tabari, Abul-Hasan Ahmad ibn Muhammad (b. Tabaristan, Persia; fl. 10th century; d. Tabaristan). Philosophy, natural science, medicine.

81. al-Tabari, Abul-Hasan Ali ibn Sahl Rabban (b. Marw, Persia, c.808; d. Baghdad, c.861). Medicine, natural science, theology, government.

82. Ibn Tariq, Yaqub (fl. Baghdad, 2nd half 8th century). Astronomy.

83. Thabit ibn Qurrah, al-Sabi al-Harrani (b. Harran, Mesopotamia [now Turkey], 836; d. Baghdad, 901). Mathematics, astronomy, mechanics, medicine, philosophy.

84. al-Tifashi, Shihab al-Din Abul-Abbas Ahmad ibn Yusuf (b. Tifash, 1184; d. Cairo, c.1254). Mineralogy, physiology.

85. Ibn al-Tilmidh, Amin al-Dawla Abul-Hasan Hibat Allah ibn Said (b. Baghdad, c.1073; d. Baghdad, 1165). Medicine, pharmacy, logic, education, literature.

86. Ibn Tufayl, Abu-Bakr Muhammad [Abubacer] (b. Guadix, Spain, c.1109; d. Marrakesh, 1185). Medicine, philosophy.

87. al-Tusi, Muhammad ibn Muhammad ibn al-Hasan [Nasir al-Din] (b. Tus, Persia, 1201; d. Kadhimain, near Baghdad, 1274). Astronomy, mathematics, mineralogy, logic, philosophy, ethics, theology. Prepared an extremely accurate table of planetary movements.

88. al-Tusi, Sharaf al-Din al-Muzaffar ibn Muhammad ibn al-Muzaffar (b. Tus; d. Persia, c.1214). Astronomy, mathematics.

89. Ulugh Beg (b. Sultaniyya, Central Asia, 1394; d. near Samarkand [now Uzbek, USSR], 1449). Astronomy.

90. Umar ibn al-Farrukhan al-Tabari (fl. Baghdad, 762-812). Astrology, astronomy.

91. al-Umawi, Abu Abdallah Yaish ibn Ibrahim ibn Yusuf ibn Simak al-Andalusi (fl. Damascus, 14th century). Arithmetic.

92. al-Uqlidisi, Abul-Hasan Ahmad ibn Ibrahim (fl. Damascus, 952-3). Arithmetic.

93. Ibn Wafid, Abu al-Mutarrif Abd al-Rahman [Abenguefit, Abenguefith, Al-benguefith, Abel Nufit] (fl. Toledo, c.1008-75). Pharmacology.

94. Ibn Wahshiyya, Abu Bakr Ahmad ibn Ali ibn al-Mukhtar (b. Qussin, near Janbala, Iraq, c.860; d. Baghdad, c.935). Agronomy, botany, alchemy, astrology, mysticism, medicine, toxicology.

95. Yahya ibn Abi Mansur (d. near Aleppo, Syria, 832). Astronomy.

96. Ibn Yaqub, Ibrahim al-Israili al-Turtushi (b. Tortosa, Spain; fl. second half 10th century). Geography.

97. Yaqubi, Ahmad ibn abi Yaqub ibn Jafar ibn Wahb ibn Wadih (fl. 9th century; d. Maghrib, 891). Geography, history.

98. Yaqut al-Hamawi al-Rumi, Shihab al-Din Abu Abdallah Yaqut ibn Abd Allah (b. Rum, Byzantine empire [now Turkey], 1179; d. Aleppo, 1229). Transmission of knowledge, geography.

99. Ibn Yunus, Abul-Hasan Ali ibn Abd al-Rahman ibn Ahmad ibn Yunus al-Sadafi (d. Fustat, Egypt, 1009). Astronomy, mathematics.

Appendix 2: Muslim Scientists

100. Ibn Yusuf, Ahmad (b. Baghdad; fl. c.900-5; d. Cairo). Mathematics.

101. al-Zahrawi, Abul-Qasim Khalaf ibn Abbas [Abulcasis] (b. al-Zahra, near Cordoba, c.936; d. al-Zahra, c.1013). Medicine, pharmacology.

102. al-Zarqali [Azarquiel], Abu Ishaq Ibrahim ibn Yahya al-Naqqash (d. Cordoba, 1100). Astronomy.

103. Ibn Zuhr, Abu Marwan Abd al-Malik ibn Abil-Ala [Abhomeron or Avenzoar] (b. Seville, c.1092; d. Seville, 1162). Medicine, toxicology, medical botany, theology.

Index

Abbasid period 66
adl (justice) 43, 44, 47, 53, 63
Adudi hospital, Baghdad 82
Africa, North 65
agricultural science 78
Alexandria 66, 78
amana 53
Andalusia 66, 76, 78
Anees, Munawar Ahmad 41-2
Arab rule 66
Arabic 66, 67
Aristotle 10, 14, 70, 71
artificial intelligence 62
Asia, Central 66
astronomy 70
Averroes, see Ibn Rushd
Avicenna, see Ibn Sina
al-Awwam, Ibn 78

Bacon, Francis 10, 13-15, 25, 105
Bacon, Roger 72-3
Baghdad 66, 82
al-Bassal, Ibn 78

al-Baytar, Abu Batr 77
Bernal, J.D. 30-1
Bhopal chemical factory 1
biologism 29
al-Biruni, Abu ar-Rayhan 66, 79, 83, 85
botany 77
Brahe, Tycho 22, 70
British Educational Reform Act (1988) 89
British Labour Party 30
British National Curriculum 90, 94
  Islamic science in 8
British Society for Social Responsibility in Science 28
Brookhaven Laboratory 18
Brown, Louise 93
Bruno, Giordano 22
Bucaille, Maurice 38-9
Butlan, al-Mukhtar bin Abdun bin 80
Byzantine culture 65

Index

Cairo 77
Callaghan, James 30
camera obscura 73-4
Carson, Rachel 1
Cartesian dualism 14
chemistry 74, 79
chemotherapy 79
Chernobyl nuclear plant 1
Christianity and science,
    hostility between 37
Christians 65
    Syrian 66
Cole, Martin 58
community orientation 51
conservation 52
contraception 58
Copernicus, Nicolaus 22, 72
Crossman, Richard 30

Damascus 72
ad-Damiri, Kamal ad-Din 77
Dark Ages of Europe 69
Darwin, Charles 13, 22, 94
    evolutionary theory 1
    revolution 22
deductivism, hypothetico 35
Demostheneses 10
Department of Health and
    Social Security (DHSS) 58
Descartes 14-15
Descartes, Rene 14
determinism, metaphysical
    (Marxist perspective) 36
dhimmis (non-Muslims living
    in Muslim state) 65
dhiya (waste) 43-4

diet 80
Dioscorides 78
diseases 80
    cure of 82
    diagnosis of 79
DNA 93
doctor-patient relationship 79
drug therapy 29
drugs 80-3

Earth 18
Easler, Brian 29
eclipse 73
    of sun, 25 May 1919 3
Eddington 3
Edinburgh 33
Edwards, Robert 93
Egypt 77, 78
Einstein's theory of relativity 4,
    16
Eliot, T.S. 63
empiricism: aprioristic, fallible
    and infallible 34
ends justify means 52
England, thirteenth-century 73
epidemiology 80
epistemological classification
    67
Euclid 73
Eudoxian model 70
eye, anatomy of 79

faith 48
falsification 35
falsificationist criteria 18
Family Planning Association 58

fard ayn 67
fard kifaya 67
Farghana 67
al-Farisi, Kamal ad-Din 73
Fars 67
al-Faruqi, Ismail 32-3
fevers 81
Feyerabend, Paul 24-5
Ford, Glyn 37
fractures 81
fragmentation 50
France 33
   Revolution 56
freedom
   from famine 1
   absolute 51
Freudenthal, Gideon 27

Gaitskell, Hugh 30
Galen 78, 81
Galileo 10, 22, 94
Gandhi, Indira 92
Gell-Mann 17
geocentrism 1, 70
geography, Islamic 85
geometry, Euclidean and
   Ptolemaic 72
al-Ghazali 67
Greece 65-6, 85
Greek language 66
green revolution 2
greenhouse effect 52
group loyalty 51

Haider, Gulzar 55
hajj 85

halal (permissible or
   praiseworthy) 43, 53
haram (forbidden or
   blameworthy) 43, 53, 54
haram (inviolate zones) 55
Harran 67
al-Haytham, Al-Hasan bin 72-4
Hayyan, Jabir Ibn 74, 76
health 80, 82
heliocentrism 71
hikma (wisdom) 63
Hippocrates 78
holistic science 50
Home Office 58
hospitals 82
human anatomy 81
Hume, David 12-13
Hunayn 78
Huygens 73
hygiene and public health 81
   Islamic 87

ibadah (worship) 43-4, 47, 63
al-Ibadi, Hunayn bin Ishaq 78
Ibn al-Haytham 69
Ibn ash-Shatir 72
Ibn Rushd 81, 87
Ibn Sina 73, 74, 80-1, 83, 87
ilm (knowledge) 43
impartiality 49
India 65, 66, 85
individualism 51
induction, principle of 13
inductivism 12
   fallible and infallible 35
industrial waste 52

Index

infectious diseases 80
information revolution 61
intellectualism, implicit 35
International Federation of
    Institutes 43
IQ testing 29
Iraq 78
Islam and science, conflict
    between 37
Islamic
  environment 53-4
  ethics and values 38
  law 65
Islamic science 7, 38, 41, 64-5,
    84
  alternative to Western
    science 6
  development of 67
  education 8
  growth of 66
  holistic 36
  immutable values of 42
  knowledge in 42
  purpose of 87
  universal 42
  value framework of 43
  what it is not 41
istihsan (preference for the
    better) 53
istislah (public interest) 43-4, 53
itidal (balance and harmony) 53

al-Jahiz 77
al-Jazzar, Ibn 80
Jeffreys 4
Jews 65, 67

judgement 50

Kant, Immanuel 16
al-Karaji 69
Kepler 22
al-Khayyami, Umar 69
khilafa (trusteeship of man)
    43-4, 47, 53, 63
Khurasan 67
al-Khwarizmi, Muhammad bin
    Musa 69
Khwarizm 67
al-Kindi 73
knowledge 25
Kufah, Iraq 74
Kuhn, Thomas 19-21, 23, 43, 91

Lavoisier 76
loyalty to God 51

Majrit (Madrid) 76
al-Majriti 76
al-Majusi, Ali bin Abbas 80-1
Makkah 84, 85
Malthus, Thomas 55-7
al-Mamun, Caliph Abdallah 70
Manzoor, S. Parvez 53
Maragha 70, 72
Marxism 28
Marxist
  historians of science 29
  view of science 29-30
Masha Allah 67
mathematics 68-9, 79
  Islamic 84
Maxwell, Nicholas 25-6

measles 79
medical education 78, 82
medicine 78, 79
    modern, development of 33
Mercury's orbit 18
Mitroff 27, 40
Moon's orbit 18
Muslim
    astronomers 70-1
    societal ethics 54

al-Nasir ad-Din 71
Nasr, Seyyed Hossein 6, 85
National Health Service 58
neutrality 51
Newton, Isaac 19, 22, 27
    his astronomy 18
    his law of gravitation 16
    his physics 27
nuclear
    spill 52
    weapons, proliferation of 28

Ockham's razor 3
optics 72
ozone layer 52

Pahari 66
paradigmism 35
paradigmism (continued)
    changes 21
    concept of 19
    Ptolemaic 21
Paris 33
partiality 49
pathology 79

persecution 67
Persia 65, 66, 76
pharmaceutical products,
    dispensing of 82
pharmacy and pharmacology
    82, 83
    Islamic 83
philosophy of knowledge 26
phrenology, social origins of 33
physics paradigm 19
physiology 81
planetary astronomy 72
pluralism, theoretical 35
poisons 81
Polanyi, Michael 28
pollution 52
Popper, Karl 15-19, 23
pornography 58
psychiatric medicine 79-80
Ptolemaic
    astronomy 70
    system 21
Ptolemy 70-1, 73

rainbow phenomenon 74
rationality 56
Ravetz, J.R. 31-2, 40
al-Razi 76
ar-Razi, Abu Bakr Muhammad
    bin Zakariya 79, 83, 87
reality, nature of 7
reductionism 50
refraction 73-4
relativism, metaphysical 35
religion and science 95-103
    in classroom 104-7

Index

religion, deliberately marginalized 5
religious morality 6
Rose, Hilary and Stephen 27, 29-30
Roszak, Theodore 28
Royal Institution, London 4
Russell, Bertrand 12, 16, 67

Sabaeans 65, 67
al-Samawal 69
Sanskrit 66
Sardar, Ziauddin 32, 34, 44, 52, 54
Sasanids 65
science
  based on Islamic thought 7
  capitalist ideology of 24
  compulsory 67
  curricula, colonial in nature 9
  dehumanizing of 24
  education, beliefs and values 89
  extraordinary 21
  form of worship 48
  in Islamic context 40
  in classroom 91, 94
  management of 51
  Marxist perspective of 30
  modern, Muslim critique of 33
  moral responsibility of 3
  no absolute certainties in 2
  non-religious 68
  normal 20
  permissible 68

science (continued)
  pleasure of God in 45
  sexist and racist 29
  socially conditioned activity 31
  value-laden 37
sciences
  life 77
  physical 68
scientific
  concept of values 23
  inquiry, irrationality of 25
  law 10
  laws, testing of 16
  research, a Pandora's box 2
  revolution 21
    in Britain 14
  statements, refutation and verification 16
  theories, limits of 3
scientists, dogmatic 2
secularization 6
Seville 78
sex 58
  crimes 58
  research 58
  Islam's attitude to 60
  morality 58
  needs and spiritual nature 59
  part of healthy body and balanced mind 59
  technologies 60
sexual revolution 61
Sijistan 67
simplicity postulate 4
Sluse 73

smallpox 79
Spain 65, 78
spices 82
spiritual crisis 52
Stalin Peace Prize 30
Steptoe, Patrick 93
sterilization 92
subjectivity 49
surgery 78, 81
synthesis 50
Syria 65, 77, 78
Syriac 66

tawheed (unity, or oneness of Allah) 43-4, 53
technological transformation 30
test-tube baby 93
Thabit bin Qurrah 67, 69
therapeutics 79
three-quark theory 18
Toledo 78

tumours 81
universalism 50
al-Urdi 70

value orientation 51
veterinary medicine 77
visual perception 72
vivisection 1

Watson, James 93
Western industrial, technological and medical progress 15
William of Ockham 3
Wilson, Harold 30

az-Zahrawi 81
zalim (tyrannical) science 44
zoological taxonomy 77
zoology 77
zulm (tyranny) 43-4

Typography by Little Red Cloud
Cover design by Frances Ross
Composed by Jasmine and Pamela Greenfield
in 11/12 pt English Times
Printed and bound by The Guernsey Press
of Vale, Guernsey, Channel Islands
on Munken 80gsm wood-free paper